Henning Mützlitz

OMEGA
HIGHLIGHTS

HEEL

HEEL Verlag GmbH
Gut Pottscheidt
53639 Königswinter
Telefon 0 22 23 / 92 30-0
Telefax 0 22 23 / 92 30 26
Mail: info@heel-verlag.de
Internet: www.heel-verlag.de

© 2009: HEEL Verlag GmbH, Königswinter

Verantwortlich für den Inhalt / Editor:
Henning Mützlitz

Alle Rechte, auch die des Nachdrucks, der Wiedergabe in jeder Form und der Übersetzung in andere Sprachen, behält sich der Herausgeber vor. Es ist ohne schriftliche Genehmigung des Verlages nicht erlaubt, das Buch und Teile daraus auf fotomechanischem Weg zu vervielfältigen oder unter Verwendung elektronischer bzw. mechanischer Systeme zu speichern, systematisch auszuwerten oder zu verbreiten.

All parts of this book are protected by copyright. Any use of this work outside the narrow limitations of copyright law is not permitted without the consent of HEEL Verlag GmbH. This applies in particular to reproduction, translation, microfilming and the saving and processing of its contents in electronic systems.

Fotos / Photos:
Archiv des Autors, Archive der Hersteller / Archive material

Englische Übersetzung / English translation:
Elizabeth Doerr

Gestaltung und Satz / Design and Layout:
Muser Medien GmbH, Mannheim
Tanja Küppershaus

Printed in Germany

– Alle Rechte vorbehalten / All rights reserved –

ISBN: 978-3-86852-196-2

INHALT

1 ....**Markengeschichte**............... Seite 6

2 ....**Historische Modelle**............... Seite 14
Eine Auswahl der schönsten Modelle aus der
Markengeschichte Omegas

3 ....**Speedmaster**............... Seite 26
Die „Moonwatch" hat sich seit 1957 ihren unverwechselbaren
Charakter bewahrt.

4 ....**„The Legend Collection"**............... Seite 40
Die Hommage an Michael Schumacher,
den größten Formel-Eins-Fahrer aller Zeiten

5 ....**Seamaster**............... Seite 44
Tief hinunter: Mit der Seamaster lassen sich
alle Winkel der Weltmeere erforschen.

6 ....**„James Bond Edition"**............... Seite 58
Auch der Geheimagent im Dienste
Ihrer Majestät schwört auf Omega.

7 ....**Chronographen**............... Seite 62
Omega hat neben Speed- und Seamaster einige
weitere Highlights im Chronographenbereich zu bieten.

8 ....**Olympia-Collection**............... Seite 72
Seit 1932 ist Omega offizieller Zeitnehmer bei den
Olympischen Spielen, so auch in Peking 2008.

9 ....**Elegante Uhren**............... Seite 80
Omega versteht sich darauf, Eleganz mit höchsten technischen
Ansprüchen zu kombinieren.

CONTENTS

1 ....**Brand history**...............Page 6

2 ....**Historical Models**...............Page 14
A selection of the most beautiful
wristwatches of Omega's brand history

3 ....**Speedmaster**...............Page 26
The Moonwatch has maintained
its unmistakable character since 1957

4 ....**The Legend Collection**...............Page 40
An homage to Michael Schumacher,
the greatest Formula 1 driver of all time

5 ....**Seamaster**...............Page 44
Down deep: the Seamaster is the perfect
companion for exploring every angle of the seven seas

6 ....**James Bond Edition**...............Page 58
Even the secret agent in the service of
Her Majesty swears by Omega

7 ....**Chronographs**...............Page 62
Alongside the Speedmaster and Seamaster models,
Omega has a great number of other chronographic highlights

8 ....**Olympic Collection**...............Page 72
Since 1932 Omega has been the official timing partner
of the Olympic Games, including Beijing 2008

9 ....**Elegant Watches**...............Page 80
Omega understands well how to combine
elegance with the highest technical demands

VORWORT

Omega ist heute der weltweit größte Uhrenhersteller und hat sowohl in Sachen Image als auch bei der technologischen Entwicklung im jungen 21. Jahrhundert Höhen erklommen, von denen in Zeiten der so genannten „Quarzkrise" in den 1970er-Jahren kaum jemand zu träumen gewagt hätte. Nicolas G. Hayek, Präsident der Swatch Group, erkannte das brachliegende Potenzial des Traditionshauses und formte innerhalb weniger Jahre eine Marke, die auf der ganzen Welt so begehrt ist wie nie zuvor in ihrer mehr als 160-jährigen Geschichte.

Wir möchten mit diesem Buch einen Einblick geben in das, was Omega heute so außergewöhnlich macht.
Neben einer Markenhistorie präsentieren wir Ihnen rund einhundert der bemerkenswertesten Modelle der Schweizer Luxusmarke. Neben den Siegern der Chronometer-Wettbewerbe vom Beginn des letzten Jahrhunderts über die legendäre „Moonwatch" bis hin zu den heutigen Co-Axial-Modellen, die den Grundstein für Omegas Zukunft als Manufaktur gelegt haben, zeigen wir Ihnen die schönsten und exklusivsten Stücke, die die Weltmarke aus Biel zu bieten hat. Dass diese Auswahl nicht vollständig sein und nur einen kleinen Teil der großen Modellvielfalt aus über 160 Jahren in Wort und Bild darstellen kann, versteht sich von selbst.
Wir wollen Sie zum Schmökern und Stöbern einladen – und vielleicht ein wenig zum Träumen.

Ein besonderer Dank geht an Christiane Hahn von Werbewelt Relations für die tatkräftige Unterstützung bei der Erstellung dieses Buches.

Heidelberg, im August 2009
Henning Mützlitz

PREFACE

Today Omega is one of the world's biggest watch manufacturers. The brand has reached a high in the beginning years of the twenty-first century in both image and technological development that nary a soul would have dared to even dream during the era of the so-called quartz crisis in the 1970s. Nicolas G. Hayek, chairman of the board of the Swatch Group, recognized the underused potential of the traditional brand and within just a few years formed a marque that is in more demand all over the world than ever before in its entire 160-year history.

With this book, we would like to provide a brief glimpse into that which makes Omega so exceptional today.
Alongside a brief history of the brand, we present you about one hundred of the most remarkable models the Swiss luxury brand manufactures. Alongside the winners of the chronometer competition at the beginning of the last century, you will see the legendary Moonwatch and the current co-axial models, which have laid the cornerstone for Omega's future as a manufacture. We will also show you some of the most beautiful and exclusive pieces that the global brand based in Biel has to offer.
It is, of course, impossible to be exhaustive here. Illustrating the entire multiplicity of models from the course of the 160-year history of Omega within these pages would be a near impossibility. Our goal, therefore, is to invite you to browse these pages and perhaps dream a little.

A big thank-you goes out to Christiane Hahn of Werbewelt Relations for her strong support in the creation of this book.

Heidelberg, August 2009
Henning Mützlitz

160 JAHRE ERFOLG

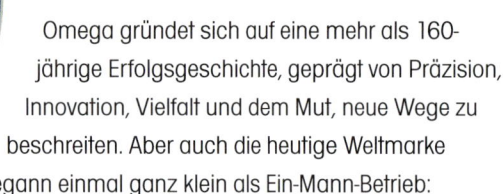

Omega-Urvater Louis Brandt
Omega founder Louis Brandt

Omega gründet sich auf eine mehr als 160-jährige Erfolgsgeschichte, geprägt von Präzision, Innovation, Vielfalt und dem Mut, neue Wege zu beschreiten. Aber auch die heutige Weltmarke begann einmal ganz klein als Ein-Mann-Betrieb: 1848 gründete der damals erst 23-jährige Louis Brandt in La Chaux-de-Fonds im schweizerischen Jura-Gebirge ein so genanntes „Comptoir d'etablissage" – einen Handwerksbetrieb, der sich der Montage von einzelnen Uhrenteilen widmete, und an den eine eigene Vertriebs- und Verkaufsabteilung angeschlossen war. Die eigenen Kreationen waren aufgrund ihres ansprechenden Äußeren schnell ein großer Erfolg. Ab dem Jahr 1877 firmierte das Unternehmen unter dem Namen „Louis Brandt & Fils", und die beiden Söhne Louis Paul und Cesar führten den Betrieb nach dem Tod des Vaters fort. 1880 zog die Firma nach Biel um, unter anderem aufgrund der besseren Verkehrsanbindung, und der Möglichkeit, kostenlose Wasserkraft für den Handwerksbetrieb nutzen zu können – in Zeiten der Industrialisierung ein immenser Standortvorteil. Die Brüder planten nämlich, künftig vollständig eigene Uhren mit selber entwickelten Uhrwerken herzustellen, eben eine komplette Uhrenmanufaktur zu etablieren. Dieser Plan wurde äußerst erfolgreich in die Tat umgesetzt, so dass ihr Betrieb bis 1889 zur größten Uhrenfabrik in der gesamten Schweiz aufstieg.

Omega-Werbeplakat anlässlich der Olympischen Spiele 1932 in Los Angeles

Omega advertising on the occasion of the Olympic Games in 1932 in Los Angeles

160 YEARS OF SUCCESS

Omega is founded on a more than 160-year history of success, characterized by precision, innovation, variety, and the courage to travel new paths.
This globally reputable brand started out as a small one-man show: in 1848, 23-year-old Louis Brandt founded a so-called comptoir d'etablissage in La Chaux-de-Fonds, a city located in Switzerland's Jura canton. This small enterprise was dedicated to the hand assembly of individual watch components and even included a department for distribution and sales. The company's own creations quickly became a great success thanks to their attractive appearances. From 1877 onward, the company was called Louis Brandt & Fils and was managed by sons Louis Paul and Cesar after the death of their father. The company moved to Biel in 1880, among other things in order to enjoy improved travel connections as well as the opportunity to use free water power— which during the era of industrialization was an immense advantage for a geographical location. The brothers had the future in mind — a future in which they were planning to completely make their own watches with movements they developed

Auch John F. Kennedy trug Omega.

John F. Kennedy also wore Omega

Auch der 14-malige Olympiasieger Michael Phelps schwört auf Omega.

Fourteen-time Olympic champion Michael Phelps swears by Omega

Der Name „Omega" wurde erstmals 1894 verwendet, als das Unternehmen mit seinem erfolgreichen Kaliber gleichen Namens einen neuen Standard hinsichtlich einfacher Konstruktion und Austauschbarkeit von Uhrwerkskomponenten setzte. 1903 wurde der Name offiziell in den Firmennamen integriert: Die „Société Anonyme Louis Brandt & Frère – Omega Watch Co" war geboren. Der Name behielt bis 1982 seine Gültigkeit, seitdem nennt sich das Unternehmen nur noch „Omega S.A.".

Genauigkeit und Präzision waren seit jeher das oberste Gebot bei Paul-Emile Brandt, der das Unternehmen von 1903 an für 50 Jahre führen sollte. 1919 gewann Omega zum ersten Mal den Genauigkeitswettbewerb des Observatoriums Neuenburg. In den 1920er- und 30er-Jahren wurden diese Erfolge mehrmals wiederholt und begründeten sogar eine eigene Uhrenlinie, die bis heute in der Kollektion von Omega einen festen Platz einnimmt: Die Constellation Chronometer, die bei den Stückzahlprüfungen der Chronometrie-Behörden lange Jahre an der Spitze lag. 1967 wurde bereits der millionste Gangschein für Omega ausgestellt. Auch heute noch zählt Omega zu den Marken, denen jedes Jahr die meisten Chronometer-Gangscheine für mechanische Armbanduhren ausgestellt werden.

themselves; in other words, they wanted to establish a complete watch manufacture. This plan was realized most successfully and saw the company becoming the largest watch factory in Switzerland by 1889.

The name Omega was first used in 1894 when the successful caliber of the same name set a new standard in terms of simple design and interchangeability of movement components. In 1903, "Omega" was officially integrated into the company's own moniker: Société Anonyme Louis Brandt & Frère – Omega Watch Co. The name remained valid until 1982; since then it has been simply called Omega SA.

Precision had always been Paul-Emile Brandt's first commandment, a man who led the watch enterprise for fifty years from 1903. In 1919, Omega won its first precision competition of the Neuchâtel observatory. In the 1920s and '30s, this success was repeated several times and even gave rise to a line that remains firmly entrenched in Omega's collection to this day: the Constellation Chronometer, which was top dog among the chronometry authorities' testing for many long years. In 1967, the one-millionth certificate was issued to Omega. And even today Omega remains one of the brands receiving the most chronometer certificates for mechanical wristwatches every year.

Another tradition began in 1909 when Omega first entered the stage as the official timing partner for a sports event, a tradition that the Swiss company has consistently maintained until

Supermodel Cindy Crawford ist seit 1995 Mitglied der „Omega-Family".

Supermodel Cindy Crawford has been a member of the "Omega family" since 1995

Eine weitere Tradition wurde 1909 begründet: Dort trat Omega erstmals offiziell als Zeitnehmer bei einer Sportveranstaltung auf, eine Tradition, die die Schweizer bis heute konsequent fortgeführt haben. Seit 1932 fungierte Omega bei mittlerweile 23 Olympischen Spielen als offizieller Zeitnehmer, das nächste Mal bei den Olympischen Winterspielen im kanadischen Vancouver im Jahr 2010. Daneben machte sich die Marke vor allem im Schwimmsport einen Namen, wo sie unzählige Male bei Weltmeisterschaften als offizieller Zeitnehmer auftrat.

Nach dem Zweiten Weltkrieg begann die Schweizer Manufaktur diejenigen Modelle zu entwickeln, die auch heute noch zur exponierten Stellung Omegas in der weltweiten Uhrenbranche beitragen: 1948 erschien die erste Seamaster und 1957 die erste Speedmaster. Vor allem die Speedmaster avancierte zum sicherlich bekanntesten und erfolgreichsten Modell der Firmengeschichte. Aufgrund seiner Zuverlässigkeit und Genauigkeit wählte die amerikanische Weltraumbehörde NASA den Chronographen 1966 zum offiziellen Ausrüstungsgegenstand. Bei der ersten Mondlandung am 21. Juli 1969 trugen die Astronauten Neil Armstrong und Buzz Aldrin die Speedmaster über ihren Raumanzügen. Bis heute war die Speedmaster die einzige Armbanduhr auf dem Mond, und Omega nennt sie seitdem stolz „Moonwatch".

Die Seamaster eroberte dagegen die Welt unter Wasser, sei es für Hobby-Taucher, die beim Schnorcheln nicht auf ihren Zeitmesser verzichten wollten, oder aber für den professionellen Einsatz bei der Erforschung der Tiefsee. Die „Ploprof" (Plongeur professionel) bewährte sich bei vielen Einsätzen industrieller Tiefseetaucher. Heute ist die Seamaster Professional eher aus Film und Fernsehen bekannt, prangt sie doch unter anderem an den Handgelenken der Hollywood-Schauspieler George Clooney und Daniel Craig.

today. Omega has been the official timer for twenty-three Olympic Games since 1932, and will once again take the stage at the 2010 Olympic Winter Games in Vancouver. Alongside this, the brand has above all made a name for itself in the discipline of swimming, where it has appeared countless times at world championships as the official timer. After World War II, this Swiss manufacture began developing the models that helped Omega to its exalted position in the global watch market: in 1948 the first Seamaster appeared, and in 1957 the first Speedmaster. It was above all the Speedmaster that advanced to become the most well known and successful model in the history of the company. Thanks to its reliability and precision, the American space agency NASA chose the chronograph to be part of its official equipment in 1966. Astronauts Neil Armstrong and Buzz Aldrin wore the Speedmaster over their space suits when they landed on the moon on July 21, 1969. To this day, the Speedmaster remains the only watch to have been on the moon, and Omega has proudly called it the Moonwatch since the momentous event occurred.

The Seamaster, on the other hand, has conquered the underwater world. It is worn both by hobby divers not wanting to leave their timekeepers on land when snorkeling and professional divers researching the deep seas. The Ploprof (Plongeur Professionel) has proven itself worthy during a great number of expeditions on the wrists of industrial deep-sea divers. Today, the Seamaster Professional is well known from television and feature films where it has been proudly worn by Hollywood actors such as George Clooney and Daniel Craig.

The development of cheap quartz watches and ensuing flood of Asian mass production on the European markets prompted a hard time for Omega at the beginning of the 1980s, characterized by great financial loss and damage to its image by desperate developments in the lower price

Die Hollywood-Schauspielerin Nicole Kidman soll die Damenwelt für die exklusiven Kreationen aus Biel begeistern.

Hollywood actor Nicole Kidman was hired to inspire women to wear the exclusive creations made in Biel

Mit der Entwicklung billiger Quarzuhren und der Überflutung des europäischen Marktes durch fernöstliche Massenware begann für Omega Anfang der 1980er-Jahre allerdings eine schwere Zeit, gekennzeichnet von hohen finanziellen Verlusten und beträchtlichen Imageschäden durch verzweifelte Entwicklungen im Niedrigpreissektor. Die Rettung kam – wie für viele andere Schweizer Traditionsmarken ebenfalls – von Nicolas G. Hayek, der seit 1983 bereits mit der Sanierung der Unternehmensgruppe ASUAG (Allgemeine Schweizerische Uhrenindustrie AG) befasst war und unter anderem den Marken Certina, Longines, Mido und Rado das Überleben sicherte. 1985 übernahm die aus der ASUAG hervorgehende SMH-Gruppe, die sich ab 1998 in Swatch Group umbenannte, die Marke Omega und integrierte sie in den heute weltweit größten Uhrenkonzern.

segment. Salvation came — as for many other traditional Swiss brands — from Nicolas G. Hayek, who reorganized the ASUAG (Allgemeine Schweizerische Uhrenindustrie AG) group, securing survival for brands such as Certina, Longines, Mido, and Rado. In 1985, the SMH Group, which emerged from the ashes of ASUAG and was renamed Swatch Group in 1998, took over Omega and integrated it into what has now become the world's largest watch concern. Hayek not only made it possible to reissue the brand's successful historical models, with an extensive image campaign he also ensured that Omega would be perceived as an exclusive lifestyle object alongside its demand on technical prowess. Actors, professional athletes, and other personalities were integrated into the "Omega family" during the 1990s. Celebrities such as Cindy Crawford, Michael

Das Kaliber 8500 mit Co-Axial-Hemmung ist das erste komplett von Omega entwickelte Uhrwerk der jüngeren Firmengeschichte.

Caliber 8500, outfitted with a co-axial escapement, is the first movement completely developed by Omega in the brand's recent history

Hayek gelang es, die alten Erfolgsmodelle neu aufzulegen und mit umfangreichen Imagekampagnen dafür zu sorgen, dass Omega neben dem technischen Anspruch vor allem als exklusives Lifestyle-Objekt wahrgenommen wurde. Schauspieler, Sportler und andere Prominente wurden im Laufe der 90er-Jahre in die „Omega-Family" integriert. Celebrities wie Cindy Crawford, Michael Schumacher, Nicole Kidman, George Clooney und in jüngerer Zeit auch der amerikanische Ausnahmeschwimmer Michael Phelps sorgten dafür, dass Omega auf der Weltbühne stetige Präsenz zeigt und allerorten Aufmerksamkeit erzeugt.

Auch in uhrmacherischer Hinsicht hat Omega in den vergangenen 15 Jahren wieder enorme Reputation erlangt: Mit dem 1994 erstmals vorgestellten Zentral-Tourbillon erklomm die Marke die Spitze der Uhrmacherkunst und mit dem vor zehn Jahren erstmals vorgestellten Co-Axial-Kaliber mit einer neuartigen, nahezu reibungsfreien Uhrwerkshemmung, die bis heute stetig weiterentwickelt und optimiert wurde, behauptet die Marke ihren Spitzenplatz in der weltweiten Uhrenindustrie.

Schumacher, Nicole Kidman, George Clooney, and — most recently — American swim star Michael Phelps ensured that Omega maintained a constant presence on the world's stage, garnering attention everywhere.

In terms of the art of watchmaking, Omega has gained enormously in reputation in the last fifteen years: the Central Tourbillon debuting in 1994 has allowed the brand to climb to the summit of haute horlogerie. The co-axial caliber first presented ten years ago with its innovative, almost friction-free escapement, has been constantly developed and optimized to this day, securing a place for the brand at the top of the global watchmaking industry.

Die Omega Speedmaster begleitete die amerikanischen Astronauten auf den Mond.

The Omega Speedmaster accompanied American astronauts to the moon

Die NASA verlieh Omega 1970 den „Snoopy Award" in Anerkennung ihrer Verdienste bei den Mondlandungen.

In 1970, NASA awarded Omega the Snoopy Award in recognition of the brand's service to the moon landing

HISTORISCHE MODELLE

Schon die ersten im Jahr 1900 hergestellten Omega-Armbanduhren wurden industriell gefertigt. Von Anfang an waren diese Uhren für ihre Genauigkeit bekannt, und nachdem viele Modelle zunächst ihren Dienst bei den Kampfeinheiten der britischen Armee in den Schützengräben des Ersten Weltkriegs versehen hatten, gewann Omega ab 1919 einen Genauigkeitswettbewerb des Observatoriums Neuenburg nach dem anderen. Seit 1957 verbindet man mit Omega vor allem die Speedmaster- und Seamaster-Modelle, die sich ihr markantes Erscheinungsbild bis heute bewahrt haben.

HISTORICAL MODELS

The first Omega wristwatches making their debut in 1900 were already industrially manufactured. Right from the beginning, these timepieces were known for their precision. Many models served in British army units in the trenches of World War I, and from 1919 Omega won one precision competition after the other at the Neuchâtel observatory. Since 1957, Omega has been particularly known for the Speedmaster and Seamaster models, models that have retained their striking appearances to this day.

TECHNICAL DATA
TECHNISCHE DATEN

CHRONOGRAPH (1915)

Referenz:	CH 568.18
Uhrwerk:	Handaufzug, Kaliber SOPB CHRO; vergoldet, grainiert und angliert; polierte Schrauben
Funktionen:	Stunden, Minuten, Kleine Sekunde; Chronograph
Gehäuse:	Silber, ø 46 mm; dreiteilig; Druckboden mit Scharnier
Bemerkung:	frühe Fliegeruhr mit 15-Minuten-Zähler und Zifferblatt aus Emaille
Schätzwert (2009):	€ 15.000,-

Reference number:	CH 568.18
Movement:	manually wound, Caliber SOPB CHRO; gold-plated, frosted and beveled; polished screws
Functions:	hours, minutes, subsidiary seconds; chronograph
Case:	silver, ø 46 mm; tripartite; push-down case back with hinged lid
Remarks:	early pilot's watch with 15-minute counter and enamel dial
Estimated value: (2009)	€ 15.000,-

HISTORICAL MODELS **15**

TECHNICAL DATA
TECHNISCHE DATEN

HERRENUHR (1927)

Uhrwerk:	Handaufzug; 15 Steine; vernickelt und geschliffen; polierte Schrauben
Funktionen:	Stunden, Minuten
Gehäuse:	Platin, 22 x 39 mm; dreiteilig; Druckboden
Bemerkung:	seltene Armbanduhr mit asymmetrischem Gehäuse
Schätzwert (2009):	€ 1100,-

MEN'S WATCH (1927)

Movement:	manually wound; 15 jewels; nickel-plated and polished; polished screws
Functions:	hours, minutes
Case:	platinum, 22 x 39 mm; tripartite; push-down case back
Remarks:	rare wristwatch with asymmetrical case
Estimated value: (2009)	€ 1100,-

SUPERLATIVE CHRONOMETER (1933)

Referenz:	2562
Uhrwerk:	Handaufzug, Kaliber 30 T2 SC Rg; geprüfter Chronometer
Funktionen:	Stunden, Minuten, Zentralsekunde
Gehäuse:	Gelbgold, ø 37 mm; Druckboden
Schätzwert (2009):	€ 2000,-

SUPERLATIVE CHRONOMETER (1933)

Reference number	2562
Movement:	manually wound, Caliber 30 T2 SC Rg; officially certified chronometer
Functions:	hours, minutes, sweep seconds
Case:	yellow gold, ø 37 mm; push-down case back
Estimated value: (2009)	€ 2000,-

COSMIC (1935)

Uhrwerk:	Handaufzug, Kaliber 27DLPC; vergoldet
Funktionen:	Stunden, Minuten, Kleine Sekunde; ewiger Kalender mit Datum, Wochentag, Monat, Mondphasen
Gehäuse:	Gelbgold, ø 34 mm; Druckboden
Schätzwert (2009):	€ 3000,-

CHRONOGRAPH (1939)

Uhrwerk:	Handaufzug, Kaliber 28.1 CHRO T1; vernickelt
Funktionen:	Stunden, Minute, Kleine Sekunde; Chronograph
Gehäuse:	Gelbgold, 32 x 40 mm; Druckboden
Schätzwert (2009):	€ 6000,-

COSMIC (1935)

Movement:	manually wound, Caliber 27DLPC; gold-plated
Functions:	hours, minutes, subsidiary seconds; perpetual calendar with date, weekday, month, moon phase
Case:	yellow gold, ø 34 mm; push-down case back
Estimated value: (2009)	€ 3000,-

CHRONOGRAPH (1939)

Movement:	manually wound, Caliber 28.1 CHRO T1; nickel-plated
Functions:	hours, minutes, subsidiary seconds; chronograph
Case:	yellow gold, 32 x 40 mm; push-down case back
Estimated value: (2009)	€ 6000,-

HISTORICAL MODELS

TECHNICAL DATA
TECHNISCHE DATEN

FLIEGERUHR ROYAL AIR FORCE (1944)

Uhrwerk:	Handaufzug, Kaliber 23.4; 15 Steine; vernickelt und geschliffen; polierte Schrauben
Funktionen:	Stunden, Minuten, Zentralsekunde
Gehäuse:	Edelstahl, ø 33 mm; dreiteilig; Lünette drehbar mit 60er-Teilung; Druckboden
Schätzwert (2009):	€ 3500,-

ROYAL AIR FORCE PILOT'S WATCH (1944)

Movement:	manually wound, Caliber 23.4; 15 jewels; nickel-plated and polished; polished screws
Functions:	hours, minutes, sweep seconds
Case:	stainless steel, ø 33 mm; tripartite; rotating bezel with 60-minute divisions; push-down case back
Estimated value: (2009)	€ 3500,-

SPEEDMASTER CHRONOGRAPH (1957)

Uhrwerk:	Handaufzug, Kaliber 321 (Lemania 2310)
Funktionen:	Stunden, Minuten, Kleine Sekunde; Chronograph
Gehäuse:	Edelstahl; ø 39 mm; Lünette mit Tachymeterskala; Hesalitglas
Bemerkung:	Die erste Speedmaster von 1957 war als erste Armbanduhr mit einer Tachymeterskala zum Messen von Geschwindigkeiten ausgestattet.
Schätzwert (2009):	€ 15.000–20.000,- (je nach Zustand)

SPEEDMASTER CHRONOGRAPH (1957)

Movement:	manually wound, Caliber 321 (Lémania 2310)
Functions:	hours, minutes, subsidiary seconds; chronograph
Case:	stainless steel; ø 46 mm; bezel with tachymeter; hesalite crystal
Remarks:	the first Speedmaster from 1957 was the first wristwatch equipped with a tachymeter to measure speeds
Estimated value: (2009)	€ 15.000–20.000,-

RAILMASTER (1956)

Uhrwerk:	Handaufzug, Kaliber 284; 17 Steine; rotvergoldet
Funktionen:	Stunden, Minuten, Zentralsekunde
Gehäuse:	Edelstahl, ø 38 mm; Boden verschraubt
Bemerkung:	antimagnetisch mit extradickem Zifferblatt
Schätzwert (2009):	€ 7000,-

RAILMASTER (1956)

Movement:	manually wound, Caliber 284; 17 jewels; red gold-plated
Functions:	hours, minutes, sweep seconds
Case:	stainless steel, ø 38 mm; screw-on case back
Remarks:	anti-magnetic with extra-thick dial
Estimated value: (2009)	€ 7000,-

HISTORICAL MODELS

TECHNICAL DATA
TECHNISCHE DATEN

RANCHERO (1957)

Uhrwerk:	Handaufzug, Kaliber 267; rotvergoldet
Funktionen:	Stunden, Minuten, Kleine Sekunde
Gehäuse:	Gelbgold, ø 35 mm; Druckboden
Schätzwert (2009):	€ 2200,-

RANCHERO (1957)

Movement:	manually wound, Caliber 267; red gold-plated
Functions:	hours, minutes, subsidiary seconds
Case:	yellow gold, ø 35 mm; push-down case back
Estimated value: (2009)	€ 2200,-

CLOISONNÉ (1963)

Uhrwerk:	Handaufzug, Kaliber 286; € 30 mm; 17 Steine; rotvergoldet; polierte Schrauben
Funktionen:	Stunden, Minuten, Zentralsekunde
Gehäuse:	Gelbgold, ø 34 mm; Druckboden
Bemerkung:	Zifferblatt aus Emaille
Schätzwert (2009):	€ 10.000,-

CLOISONNÉ (1963)

Movement:	manually wound; Caliber 286; ø 30 mm; 17 jewels; red gold-plated; polished screws
Functions:	hours, minutes, sweep seconds
Case:	platinum, 22 x 39 mm; push-down case back
Remarks:	enamel dial
Estimated value: (2009)	€ 10.000,-

HISTORISCHE MODELLE

DE VILLE CHRONOGRAPH (1967)

Uhrwerk:	Handaufzug, Kaliber 320; vergoldet
Funktionen:	Stunden, Minuten, Kleine Sekunde; Chronograph
Gehäuse:	Gelbgold, ø 35 mm; Druckboden
Schätzwert (2009):	€ 2500,-

DE VILLE CHRONOGRAPH (1967)

Movement:	manually wound, Caliber 320; gold-plated
Functions:	hours, minutes, subsidiary seconds; chronograph
Case:	yellow gold, ø 35 mm; push-down case back
Estimated value: (2009)	€ 2500,-

SPEEDMASTER PROFESSIONAL CHRONOGRAPH (1968)

Uhrwerk:	Handaufzug, Kaliber 321; rotvergoldet
Funktionen:	Stunden, Minuten, Kleine Sekunde; Chronograph
Gehäuse:	Edelstahl, ø 42 mm; Boden mit Sichtfenster
Schätzwert (2009):	€ 2500,-

SPEEDMASTER PROFESSIONAL CHRONOGRAPH (1968)

Movement:	manually wound, Caliber 321; red gold-plated
Functions:	hours, minutes, subsidiary seconds; chronograph
Case:	stainless steel, ø 42; transparent case back
Estimated value: (2009)	€ 2500,-

HISTORICAL MODELS

TECHNICAL DATA
TECHNISCHE DATEN

SEAMASTER MEMOMATIC (1969)

Uhrwerk:	Automatik, Kaliber 980; rotvergoldet
Funktionen:	Stunden, Minuten, Zentralsekunde; Datum; mechanischer Wecker
Gehäuse:	Edelstahl, 40 x 43 mm; Boden verschraubt
Bemerkung:	Armbandwecker mit minutengenauer Einstellung der Weckzeit
Schätzwert (2009):	€ 1300,-

SEAMASTER MEMOMATIC (1969)

Movement:	automatic, Caliber 980; red gold-plated
Functions:	hours, minutes, sweep seconds; date; mechanical alarm
Case:	stainless steel, 40 x 43 mm; screw-down case back
Remarks:	wristwatch with alarm that can be set to the minute
Estimated value: (2009)	€ 1300,-

SEAMASTER CHRONOSTOP (1970)

Uhrwerk:	Handaufzug, Kaliber 865; rotvergoldet
Funktionen:	Stunden, Minuten; Chronograph
Gehäuse:	Edelstahl, 41 x 47 mm; Zifferblatt-Drehring mit 24er-Teilung; Boden verschraubt
Schätzwert (2009):	€ 2000,-

SEAMASTER CHRONOSTOP (1970)

Movement:	manually wound, Caliber 865; 15 jewels; rose gold-plated
Functions:	hours, minutes; chronograph
Case:	stainless steel, 41 x 47 mm; rotating ring with 24-hour divisions; screw-down case back
Estimated value: (2009)	€ 2000,-

HISTORISCHE MODELLE

AUTOMATIC CHRONOGRAPH „FIFA" (1970)

Uhrwerk:	Automatik, Kaliber 1045; rotvergoldet
Funktionen:	Stunden, Minuten, Kleine Sekunde; Chronograph
Gehäuse:	Edelstahl, 40 x 43 mm; Boden verschraubt
Bemerkung:	Sondermodell für den Fußball-Weltverband FIFA
Schätzwert (2009):	€ 2600,-

AUTOMATIC CHRONOGRAPH FIFA (1970)

Movement:	automatic, Caliber 1045; red gold-plated
Functions:	hours, minutes, subsidiary seconds; chronograph
Case:	stainless steel, 40 x 43 mm; screw-down case back
Remarks:	limited edition for the International Soccer Association FIFA
Estimated value: (2009)	€ 2600,-

SPEEDMASTER CHRONOGRAPH AUTOMATIC (1978)

Uhrwerk:	Automatik, Kaliber 1045; rotvergoldet
Funktionen:	Stunden, Minuten, Kleine Sekunde; Chronograph; Datum, Wochentag, 24-Std.-Anzeige (zweite Zeitzone)
Gehäuse:	Edelstahl, ø 40 mm; Boden verschraubt
Schätzwert (2009):	€ 1100,-

SPEEDMASTER CHRONOGRAPH AUTOMATIC (1978)

Movement:	automatic, Caliber 1045; red gold-plated
Functions:	hours, minutes, subsidiary seconds; chronograph; date, weekday; 24-hour display (second time zone)
Case:	stainless steel; ø 40 mm; screw-down case back
Estimated value: (2009)	€ 1100,-

TECHNICAL DATA
TECHNISCHE DATEN

SEAMASTER CHRONOGRAPH „DARTH VADER" (1970)

Uhrwerk:	Handaufzug, Kaliber 861; rotvergoldet
Funktionen:	Stunden, Minuten, Kleine Sekunde; Chronograph
Gehäuse:	Edelstahl, geschwärzt, 44 x 46 mm; Boden verschraubt
Schätzwert (2009):	€ 3500,-

SEAMASTER CHRONOGRAPH DARTH VADER (1970)

Movement:	manually wound, Caliber 861; red gold-plated
Functions:	hours, minutes, subsidiary seconds; chronograph
Case:	blackened stainless steel, 44 x 46 mm; screw-down case back
Estimated value: (2009)	€ 3500,-

SPEEDMASTER PROFESSIONAL „ANAKIN SKYWALKER" (1973)

Uhrwerk:	Handaufzug, Kaliber 861; rotvergoldet
Funktionen:	Stunden, Minuten, Kleine Sekunde; Chronograph
Gehäuse:	Edelstahl, 45 x 51 mm; Boden verschraubt
Schätzwert (2009):	€ 3500,-

SPEEDMASTER PROFESSIONAL ANAKIN SKYWALKER (1973)

Movement:	manually wound; Caliber 861; red gold-plated
Functions:	hours, minutes; chronograph; date; 24-hour display
Case:	stainless steel, 42 x 51 mm; screw-down case back
Estimated value: (2009)	€ 3500,-

SPEEDMASTER 125 (1977)

Uhrwerk:	Automatik, Kaliber 1041; 22 Steine; rotvergoldet
Funktionen:	Stunden, Minuten; Chronograph; Datum; 24-Std.-Anzeige
Gehäuse:	Edelstahl, 42 x 51 mm; Boden verschraubt
Bemerkung:	Jubiläumsmodell zum 125-jährigen Firmenjubiläum
Schätzwert (2009):	€ 1600,-

SPEEDMASTER 125 (1977)

Movement:	automatic, Caliber 1041; 22 jewels; red gold-plated
Functions:	hours, minutes; chronograph; date; 24-hour display
Case:	stainless steel, 42 x 51 mm; screw-on case back
Remarks:	anniversary model on the occasion of the company's 125th anniversary
Estimated value: (2009)	€ 1600,-

SEAMASTER AUTOMATIC 600M/2000 FT PROFESSIONAL („PLOPROF") (1972)

Uhrwerk:	Automatik, Kaliber 1002; rotvergoldet
Funktionen:	Stunden, Minuten, Zentralsekunde; Datum
Gehäuse:	Edelstahl, 56 x 45 mm; Monocoque-Konstruktion; Lünette drehbar mit 60er-Teilung und Sicherheitsdrücker; wasserdicht bis 60 bar (600 m)
Bemerkung:	Professionelle Taucheruhr für den Tiefseeeinsatz
Schätzwert (2009):	€ 6500,-

SEAMASTER AUTOMATIC 600M/2000 FT PROFESSIONAL PLOPROF (1972)

Movement:	manually wound, Caliber 1002; red gold-plated
Functions:	hours, minutes, sweep seconds; date
Case:	stainless steel, 56 x 45 mm; monocoque construction; rotating bezel with 60-minute divisions and security buttons; water-resistant to 60 atm (600 m)
Remarks:	professional deep-sea diver's watch
Estimated value: (2009)	€ 6500,-

HISTORICAL MODELS

SPEEDMASTER

Kaum eine andere Uhr wird so sehr mit der Marke Omega in Verbindung gebracht wie die Speedmaster. Die Lancierung der ersten Speedmaster im Jahr 1957 fiel genau in die Pionierzeit der Raumfahrt, und der robuste Chronograph wurde 1965 von der amerikanischen Weltraumbehörde NASA ausgewählt, um ihre Astronauten auf den riskanten Missionen zum Mond zu begleiten. Bis 1972 war sie bei allen bemannten Mond-Missionen dabei, und bis heute ist die Speedmaster die einzige Uhr, die jemals auf dem Erdtrabanten gewesen ist – aus diesem Grund nennt Omega sie „Moonwatch" und verweist seit nunmehr vier Jahrzehnten immer wieder mit Sondermodellen auf die einzigartige Historie dieses besonderen Chronographen.

There is hardly another watch so tied to the brand name Omega as the Speedmaster. The launch of the first Speedmaster in 1957 took place exactly as space travel was in the pioneering phase, and the robust chronograph was chosen in 1965 by the American space agency NASA to accompany its astronauts on the risky mission to the moon. It took part in all manned moon missions until 1972, and to this very day the Speedmaster remains the only watch ever to have been on the earth's satellite. For this reason, it is called the Moonwatch and for four decades special edition models have continued to allude to the unique history of this special chronograph.

TECHNICAL DATA
TECHNISCHE DATEN

SPEEDMASTER PROFESSIONAL MOONWATCH

Referenz:	3570.50.00
Uhrwerk:	Handaufzug, Omega Kaliber 1861
Funktionen:	Stunden, Minuten, Kleine Sekunde; Chronograph
Gehäuse:	Edelstahl, ø 42 mm; Lünette mit Tachymeterskala; Saphirglas; wasserdicht bis 5 bar (50 m)
Preis (2006):	€ 2320,-

SPEEDMASTER PROFESSIONAL MOONWATCH

Reference number:	3570.50.00
Movement:	manually wound, Omega Caliber 1861
Functions:	hours, minutes, subsidiary seconds; chronograph
Case:	stainless steel, ø 42 mm; bezel with tachymeter; sapphire crystal; water-resistant to 5 atm (50 m)
Price (2006):	€ 2320,-

SPEEDMASTER MOONWATCH CO-AXIAL

Referenz:	311.30.44.50.01.002
Uhrwerk:	Automatik, Omega Kaliber 3313; Co-Axial-Hemmung; geprüfter Chronometer (COSC)
Funktionen:	Stunden, Minuten, Kleine Sekunde; Chronograph; Datum
Gehäuse:	Edelstahl, ø 44,35 mm; Lünette mit Tachymeterskala; Saphirglas; Boden mit Sichtfenster; wasserdicht bis 10 bar (100 m)
Preis (2008):	€ 4300,-

SPEEDMASTER MOONWATCH CO-AXIAL

Reference number:	311.30.44.50.01.002
Movement:	automatic, Omega Caliber 3313; co-axial escapement; officially certified chronometer (COSC)
Functions:	hours, minutes, subsidiary seconds; chronograph; date
Case:	stainless steel, ø 44.35 mm; bezel with tachymeter; sapphire crystal; transparent case back; water-resistant to 10 atm (100 m)
Price (2008):	€ 4300,-

SPEEDMASTER PROFESSIONAL MOONWATCH „SKYLAB 3"

Referenz:	3597.23
Uhrwerk:	Handaufzug, Omega Kaliber 1861
Funktionen:	Stunden, Minuten, Kleine Sekunde; Chronograph
Gehäuse:	Edelstahl, ø 42 mm; Lünette mit Tachymeterskala; Saphirglas; wasserdicht bis 5 bar (50 m)
Preis:	€ 1942,-

SPEEDMASTER PROFESSIONAL MOONWATCH SKYLAB 3

Reference number:	3597.23
Movement:	manually wound, Omega Caliber 1861
Functions:	hours, minutes, subsidiary seconds; chronograph
Case:	stainless steel, ø 42 mm; bezel with tachymeter; sapphire crystal; water-resistant to 5 atm (50 m)
Price:	€ 1942,-

SPEEDMASTER PROFESSIONAL MOONWATCH „APOLLO XIII"

Referenz:	3595.52.00
Uhrwerk:	Handaufzug, Omega Kaliber 1861
Funktionen:	Stunden, Minuten, Kleine Sekunde; Chronograph
Gehäuse:	Edelstahl, ø 42 mm; Lünette mit Tachymeterskala; Saphirglas; wasserdicht bis 5 bar (50 m)
Preis:	n. b.

SPEEDMASTER PROFESSIONAL MOONWATCH APOLLO XIII

Reference number:	3595.52.00
Movement:	manually wound, Omega Caliber 1861
Functions:	hours, minutes, subsidiary seconds; chronograph
Case:	stainless steel, ø 42 mm; bezel with tachymeter; sapphire crystal; water-resistant to 5 atm (50 m)
Price:	u/a

SPEEDMASTER PROFESSIONAL MOONWATCH „SNOOPY AWARD" LIMITED EDITION

Referenz:	3578.51.00
Uhrwerk:	Handaufzug, Omega Kaliber 1861
Funktionen:	Stunden, Minuten, Kleine Sekunde; Chronograph
Gehäuse:	Edelstahl, ø 42 mm; Lünette mit Tachymeterskala; Hesalitglas; wasserdicht bis 5 bar (50 m)
Preis (2003):	€ 2540,-

SPEEDMASTER PROFESSIONAL MOONWATCH SNOOPY AWARD LIMITED EDITION

Reference number:	3578.51.00
Movement:	manually wound, Omega Caliber 1861
Functions:	hours, minutes, subsidiary seconds; chronograph
Case:	stainless steel, ø 42 mm; bezel with tachymeter; hesalite crystal; water-resistant to 5 atm (50 m)
Price (2003):	€ 2540,-

TECHNICAL DATA
TECHNISCHE DATEN

SPEEDMASTER PROFESSIONAL MOONWATCH „FROM MOON TO MARS"

Referenz:	3577.50.00
Uhrwerk:	Handaufzug, Omega Kaliber 1861
Funktionen:	Stunden, Minuten, Kleine Sekunde; Chronograph
Gehäuse:	Edelstahl, ø 42 mm; Lünette mit Tachymeterskala; Saphirglas; wasserdicht bis 5 bar (50 m)
Preis (2004):	€ 2920,-

SPEEDMASTER PROFESSIONAL MOONWATCH FROM MOON TO MARS

Reference number:	3577.50.00
Movement:	manually wound, Omega Caliber 1861
Functions:	hours, minutes, subsidiary seconds; chronograph
Case:	stainless steel, ø 42 mm; bezel with tachymeter; hesalite crystal; water-resistant to 5 atm (50 m)
Price (2004):	€ 2920,-

SPEEDMASTER PROFESSIONAL MOONWATCH „APOLLO 11 40TH ANNIVERSARY" LIMITED EDITION

Referenz:	311.90.42.30.01.001
Uhrwerk:	Handaufzug, Omega Kaliber 1861
Funktionen:	Stunden, Minuten, Kleine Sekunde; Chronograph
Gehäuse:	Platin, ø 42 mm; Lünette mit Tachymeterskala; Hesalitglas; wasserdicht bis 5 bar (50 m)
Bemerkung:	Medaille auf Zifferblatt und Missionspatch auf Gehäuseboden in Gelbgold
Preis (2009):	€ 67.560,-

SPEEDMASTER PROFESSIONAL MOONWATCH APOLLO 11 40TH ANNIVERSARY LIMITED EDITION

Reference number:	311.90.42.30.01.001
Movement:	manually wound, Omega Caliber 1861
Functions:	hours, minutes, subsidiary seconds; chronograph
Case:	platinum, ø 42 mm; bezel with tachymeter; hesalite crystal; water-resistant to 5 atm (50 m)
Remarks:	medal on dial and mission patch on case back in yellow gold
Price (2009):	€ 67.560,-

TECHNICAL DATA
TECHNISCHE DATEN

SPEEDMASTER „50TH ANNIVERSARY PATCH" LIMITED EDITION

Referenz:	311.30.42.30.01.001
Uhrwerk:	Handaufzug, Omega Kaliber 1861
Funktionen:	Stunden, Minuten, Kleine Sekunde; Chronograph
Gehäuse:	Edelstahl, ø 42 mm; Lünette mit Tachymeterskala; Saphirglas; wasserdicht bis 5 bar (50 m)
Preis (2007):	€ 3130,-

SPEEDMASTER 50TH ANNIVERSARY PATCH LIMITED EDITION

Reference number:	311.30.42.30.01.001
Movement:	manually wound, Omega Caliber 1861
Functions:	hours, minutes, subsidiary seconds; chronograph
Case:	stainless steel, ø 42 mm; bezel with tachymeter; sapphire crystal, water-resistant to 5 atm (50 m)
Price (2007):	€ 3130,-

SPEEDMASTER PROFESSIONAL MOONWATCH CO-AXIAL „50TH ANNIVERSARY" LIMITED EDITION

Referenz:	311.63.42.50.01.001 (Roségold), 311.63.42.50.01.002 (Gelbgold)
Uhrwerk:	Handaufzug, Omega Kaliber 3201; Co-Axial-Hemmung; geprüfter Chronometer (COSC)
Funktionen:	Stunden, Minuten, Kleine Sekunde; Chronogaph
Gehäuse:	Roségold / Gelbgold, ø 42 mm; Lünette mit Tachymeterskala; Saphirglas; Boden mit Sichtfenster; wasserdicht bis 10 bar (100 m)
Preis (2007):	je € 21.360,-

SPEEDMASTER PROFESSIONAL MOONWATCH CO-AXIAL 50TH ANNIVERSARY LIMITED EDITION

Reference number:	311.63.42.50.01.001 (rose gold), 311.63.42.50.01.002 (yellow gold)
Movement:	manually wound, Omega Caliber 3201; co-axial escapement; officially certified chronometer (COSC)
Functions:	hours, minutes, subsidiary seconds; chronograph
Case:	rose gold/yellow gold, ø 42 mm; bezel with tachymeter; sapphire crystal; transparent case back; water-resistant to 10 atm (100 m)
Price (2007):	€ 21.360,-

SPEEDMASTER BROAD ARROW CO-AXIAL 1957

Referenz:	321.93.42.50.02.001
Uhrwerk:	Automatik, Omega Kaliber 3313; Co-Axial-Hemmung; geprüfter Chronometer (COSC)
Funktionen:	Stunden, Minuten, Kleine Sekunde; Chronograph; Datum
Gehäuse:	Edelstahl, ø 42 mm; Lünette in Gelbgold mit Tachymeterskala; Saphirglas; Boden mit Sichtfenster; wasserdicht bis 10 bar (100 m)
Preis (2007):	€ 5340,-

SPEEDMASTER BROAD ARROW CO-AXIAL 1957

Reference number:	321.93.42.50.02.001
Movement:	automatic, Omega Caliber 3313; co-axial escapement; officially certified chronometer (COSC)
Functions:	hours, minutes, subsidiary seconds; chronograph; date
Case:	stainless steel, ø 42 mm; yellow-gold bezel with tachymeter; sapphire crystal; transparent case back; water-resistant to 10 atm (100 m)
Price (2007):	€ 5340,-

TECHNICAL DATA
TECHNISCHE DATEN

**SPEEDMASTER
BROAD ARROW CO-AXIAL 1957**

Referenz:	321.53.42.50.02.001 (Roségold), 321.50.42.50.02.001 (Gelbgold)
Uhrwerk:	Automatik, Omega Kaliber 3313; Co-Axial-Hemmung; geprüfter Chronometer (COSC)
Funktionen:	Stunden, Minuten, Kleine Sekunde; Chronograph; Datum
Gehäuse:	Roségold / Gelbgold, ø 42 mm; Lünette mit Tachymeterskala; Saphirglas; Boden mit Sichtfenster; wasserdicht bis 10 bar (100 m)
Preis (2007):	je € 9920,-

SPEEDMASTER BROAD ARROW CO-AXIAL 1957

Reference number:	321.53.42.50.02.001 (rose gold), 321.50.42.50.02.001 (yellow gold)
Movement:	automatic, Omega Caliber 3313; co-axial escapement; officially certified chronometer (COSC)
Functions:	hours, minutes, subsidiary seconds; chronograph; date
Case:	rose gold/yellow gold, ø 42 mm; bezel with tachymeter; sapphire crystal; transparent case back; water-resistant to 10 atm (100 m)
Price (2007):	€ 9920,-

SPEEDMASTER

TECHNICAL DATA
TECHNISCHE DATEN

SPEEDMASTER CO-AXIAL
BROAD ARROW CHRONOGRAPH RATTRAPANTE

Referenz:	3882.31.37
Uhrwerk:	Automatik, Omega Kaliber 3612; Co-Axial-Hemmung; geprüfter Chronometer (COSC)
Funktionen:	Stunden, Minuten, Kleine Sekunde; Schleppzeiger-Chronograph; Datum
Gehäuse:	Edelstahl, ø 44,25 mm; Lünette mit Tachymeterskala; Saphirglas
Preis (2006):	€ 8920,-

SPEEDMASTER CO-AXIAL
BROAD ARROW CHRONOGRAPH RATTRAPANTE

Reference number:	3882.31.37
Movement:	automatic, Omega Caliber 3612; co-axial escapement; officially certified chronometer (COSC)
Functions:	hours, minutes, subsidiary seconds; flyback chronograph; date
Case:	stainless steel, ø 44.25 mm; bezel with tachymeter; sapphire crystal
Price (2006):	€ 8920,-

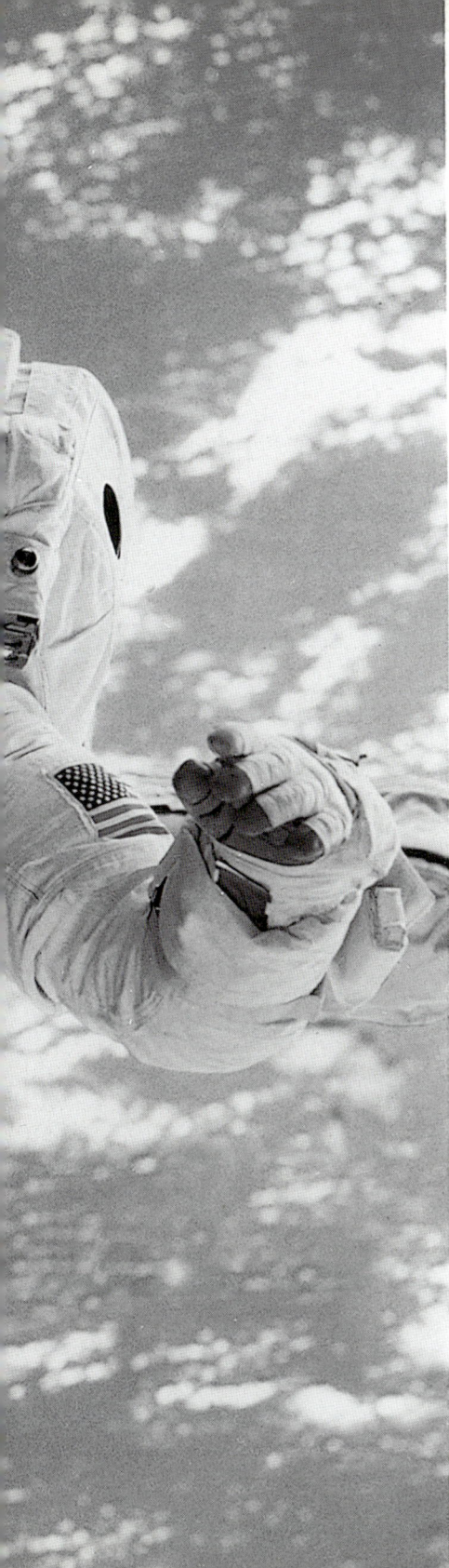

SPEEDMASTER BROAD ARROW SKELETON

Referenz:	178.00.24
Uhrwerk:	Handaufzug, Omega Kaliber 3221; mit Genfer Streifenschliff; Rotor aus Platin; geprüfter Chronometer (COSC)
Funktionen:	Stunden, Minuten, Kleine Sekunde; Chronograph
Gehäuse:	Platin, ø 42 mm; Lünette mit Tachymeterskala; Saphirglas; wasserdicht bis 10 bar (100 m)
Preis (2000):	€ 43.770,-

SPEEDMASTER BROAD ARROW SKELETON

Reference number:	178.00.24
Movement:	manually wound, Omega Caliber 3221; côtes de Genève; platinum rotor; officially certified chronometer (COSC)
Functions:	hours, minutes, subsidiary seconds; chronograph
Case:	platinum, ø 42 mm; bezel with tachymeter; sapphire crystal; water-resistant to 10 atm (100 m)
Price (2000):	€ 43.770,-

SPEEDMASTER PROFESSIONAL X-33

Referenz:	3991.50.41
Uhrwerk:	quarzgesteuert, Omega Kaliber 1666; 24 Monate Batterielaufzeit
Funktionen:	Stunden, Minuten, Zentralsekunde; Chronograph; Universalzeit; Alarmfunktion
Gehäuse:	Titan, ø 41 mm; Lünette einseitig drehbar mit 60er-Teilung; Saphirglas; wasserdicht bis 3 bar (30 m)
Preis (2002):	€ 2260,-

SPEEDMASTER PROFESSIONAL X-33

Reference number:	3991.50.41
Movement:	quartz, Omega Caliber 1666; 24-month battery life
Functions:	hours, minutes, sweep seconds; chronograph; universal time; alarm function
Case:	titanium, ø 41 mm; unidirectionally rotating bezel with 60-minute divisions; sapphire crystal; transparent case back; water-resistant to 3 atm (30 m)
Price (2002):	€ 2260,-

TECHNICAL DATA
TECHNISCHE DATEN

SPEEDMASTER PROFESSIONAL ALASKA PROJECT LIMITED EDITION

Referenz:	311.32.42.30.04.001
Uhrwerk:	Handaufzug, Omega Kaliber 1861
Funktionen:	Stunden, Minuten, Kleine Sekunde; Chronograph
Gehäuse:	Edelstahl, ø 41 mm (58 mm); Lünette aus Aluminium mit 60er-Teilung; Saphirglas; wasserdicht bis 5 bar; Temperatur-resistent von -148° bis +260° Grad Celsius
Bemerkung:	wärmeisolierendes Übergehäuse mit Übertragungsdrückern abnehmbar
Preis (2008):	€ 4230,-

  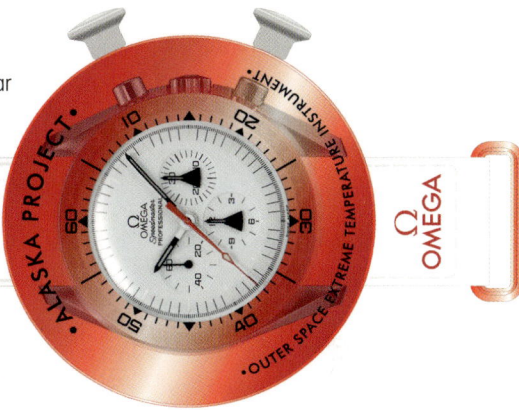

SPEEDMASTER PROFESSIONAL ALASKA PROJECT LIMITED EDITION

Reference number:	311.32.42.30.04.001
Movement:	manually wound, Omega Caliber 1861
Functions:	hours, minutes, subsidiary seconds; chronograph
Case:	stainless steel, ø 41 mm (58 mm); aluminum bezel with 60-minute divisions; sapphire crystal; water-resistant to 5 atm; temperature-resistant from -148° to +260° Celsius
Remarks:	heat-isolating outer case outfitted with removable transmission button
Price (2008):	€ 4230,-

SPEEDMASTER „SOLAR IMPULSE" CO-AXIAL GMT

Referenz:	321.92.44.52.01.001
Uhrwerk:	Automatik, Omega Kaliber 3603; Co-Axial-Hemmung; geprüfter Chronometer
Funktionen:	Stunden, Minuten, Kleine Sekunde; Chronograph; Datum; 24-Std.-Anzeige (zweite Zeitzone)
Gehäuse:	Titan, ø 44,25 mm; Lünette mit Tachymeterskala; Saphirglas; wasserdicht bis 10 bar
Preis (2007):	€ 5690,-

SPEEDMASTER SOLAR IMPULSE CO-AXIAL GMT

Reference number:	321.92.44.52.01.001
Movement:	automatic, Omega Caliber 3603; co-axial escapement; officially certified chronometer
Functions:	hours, minutes, subsidiary seconds; chronograph; date
Case:	titanium, ø 44.25 mm; bezel with tachymeter; sapphire crystal; water-resistant to 10 atm (100 m)
Price (2007):	€ 5690,-

SPEEDMASTER

OMEGA
„THE LEGEND COLLECTION"

Rennfahrer sind seit jeher begehrt, wenn es darum geht, Marken oder Produkte zu präsentieren. Während andere froh sind, überhaupt irgendeinen Fahrer unterstützen zu dürfen, ging Omega keine Kompromisse ein: Mit Michael Schumacher versicherte man sich kurzerhand der Dienste des mit Abstand erfolgreichsten Formel-Eins-Fahrers der Rennhistorie. Seit 1995 ist der siebenmalige Weltmeister Teil der Omega-Familie. In dieser Zeit wirkte er immer wieder selbst bei der Gestaltung der verschiedenen Speedmaster-Modelle seiner „Legend-Collection" mit.

Race car drivers have always been greatly sought-after when it comes to endorsing brands and products. While some enterprises can be satisfied just to get any driver to work with them, Omega has never stood for compromise: in the person of Michael Schumacher they were assured the services of the most successful race car driver in Formula 1 history. Since 1995, the seven-time world champion has been part of the "Omega family." During this time, he himself has taken part in the design of various Speedmaster models of his Legend Collection.

TECHNICAL DATA
TECHNISCHE DATEN

SPEEDMASTER NEW DATE

Referenz:	3210.51.00
Uhrwerk:	Automatik, Omega Kaliber 1164; geprüfter Chronometer (COSC)
Funktionen:	Stunden, Minuten, Kleine Sekunde; Chronograph, Datum
Gehäuse:	Edelstahl, ø 40 mm; Lünette mit Tachymeterskala; Saphirglas; wasserdicht bis 10 bar (100 m)
Preis (2006):	€ 2175,-

SPEEDMASTER NEW DATE

Reference number:	3210.51.00
Movement:	automatic, Omega Caliber 1164; officially certified chronometer (COSC)
Functions:	hours, minutes, subsidiary seconds; chronograph, date
Case:	stainless steel, ø 40 mm; bezel with tachymeter; sapphire crystal; water-resistant to 10 atm (100 m)
Price (2006):	€ 2175,-

THE LEGEND COLLECTION

SPEEDMASTER LEGEND

Referenz:	3506.61.00
Uhrwerk:	Automatik, Omega Kaliber 3301; geprüfter Chronometer (COSC)
Funktionen:	Stunden, Minuten, Kleine Sekunde; Chronograph, Datum
Gehäuse:	Edelstahl, ø 42 mm; Lünette mit Tachymeterskala; Saphirglas; wasserdicht bis 10 bar (100 m)
Preis (2006):	€ 3410,-

SPEEDMASTER LEGEND

Reference number:	3506.61.00
Movement:	automatic, Omega Caliber 3301; officially certified chronometer (COSC)
Functions:	hours, minutes, subsidiary seconds; chronograph, date
Case:	stainless steel, ø 42 mm; bezel with tachymeter; sapphire crystal; water-resistant to 10 atm (100 m)
Price (2006):	€ 3410,-

SPEEDMASTER BROAD ARROW „MICHAEL SCHUMACHER" LIMITED EDITION

Referenz:	3159.50.00
Uhrwerk:	Automatik, Omega Kaliber 3303; geprüfter Chronometer (COSC)
Funktionen:	Stunden, Minuten, Kleine Sekunde; Chronograph; Datum
Gehäuse:	Rotgold, ø 40 mm; Lünette mit Tachymeterskala; Saphirglas; wasserdicht bis 10 bar (100 m)
Preis (2003):	€ 15.900,-

SPEEDMASTER BROAD ARROW MICHAEL SCHUMACHER LIMITED EDITION

Reference number:	3159.50.00
Movement:	automatic, Omega Caliber 3303; officially certified chronometer (COSC)
Functions:	hours, minutes, subsidiary seconds; chronograph; date
Case:	red gold, ø 40 mm; bezel with tachymeter; sapphire crystal; water-resistant to 10 atm (100 m)
Price (2003):	€ 15.900,-

THE LEGEND COLLECTION

TECHNICAL DATA
TECHNISCHE DATEN

SPEEDMASTER „THE LEGEND" COLLECTION
CO-AXIAL 2008

Referenz:	321.30.44.50.01.001
Uhrwerk:	Automatik, Omega Kaliber 3313; Co-Axial-Hemmung; geprüfter Chronometer (COSC)
Funktionen:	Stunden, Minuten, Kleine Sekunde; Chronograph; Datum
Gehäuse:	Edelstahl, ø 44,25 mm; Lünette mit dem Palmarès von Michael Schumacher; Saphirglas; wasserdicht bis 10 bar (100 m)
Preis (2008):	€ 4230,-

SPEEDMASTER THE LEGEND COLLECTION
CO-AXIAL 2008

Reference number:	321.30.44.50.01.001
Movement:	automatic, Omega Caliber 3313; co-axial escapement; officially certified chronometer (COSC)
Functions:	hours, minutes, subsidiary seconds; chronograph; date
Case:	stainless steel, ø 44.25 mm; bezel lists Michael Schumacher's accomplishments; sapphire crystal; water-resistant to 10 atm (100 m)
Price (2008):	€ 4230,-

THE LEGEND COLLECTION

OMEGA
SEAMASTER

Neben der allgegenwärtigen Speedmaster hat sich vor allem die Seamaster in den vergangenen Jahren zum Aushängeschild von Omega entwickelt. Die 1948 erstmals vorgestellte Taucheruhr ist mittlerweile in unzähligen Variationen erhältlich: Vom professionellen Einsatzinstrument in der Tiefsee (Seamaster „Ploprof") über die klassische sportlich-elegante Taucheruhr bis hin zu den bunten „Planet Ocean"-Modellen, die am Handgelenk aufgrund ihrer schrillen Farben oder dem mitunter opulenten Edelsteinbesatz zum Hingucker werden – hier ist für jeden etwas geboten!

Along with the Speedmaster it is above all the Seamaster that has developed into Omega's flagship in the last few years. The diver's watch first presented in 1948 is meanwhile available in countless incarnations: a professional instrument for deep seas (Seamaster PloProf); classic sporty, elegant diver's watch; and a colorful variation in the Planet Ocean models, which are total eye-catchers on the wrist thanks to their shrill colors and/or opulent gemstones. This collection truly offers something for every taste.

TECHNICAL DATA
TECHNISCHE DATEN

SEAMASTER PLANET OCEAN CO-AXIAL

Referenz:	2909.50.38.f.c.
Uhrwerk:	Automatik, Omega Kaliber 2500 C; Co-Axial-Hemmung; geprüfter Chronometer (COSC)
Funktionen:	Stunden, Minuten, Zentralsekunde; Datum
Gehäuse:	Edelstahl, ø 42 mm; Lünette einseitig drehbar mit 60er-Teilung; Saphirglas; Krone verschraubt; Heliumventil; wasserdicht bis 60 bar (600 Meter)
Preis (2006):	€ 2670,-

SEAMASTER PLANET OCEAN CO-AXIAL

Reference number:	2909.50.38.f.c.
Movement:	automatic, Omega Caliber 2500 C; co-axial escapement; officially certified chronometer (COSC)
Functions:	hours, minutes, sweep seconds; date
Case:	stainless steel, ø 42 mm; unidirectionally rotating bezel with 60-minute divisions; sapphire crystal; screw-in crown; helium valve; water-resistant to 60 atm (600 m)
Price (2006):	€ 2670,-

SEAMASTER PLANET OCEAN CO-AXIAL CHRONOGRAPH

Referenz:	2918.50.38
Uhrwerk:	Automatik, Omega Kaliber 3313; Co-Axial-Hemmung; geprüfter Chronometer (COSC)
Funktionen:	Stunden, Minuten, Kleine Sekunde; Chronograph; Datum
Gehäuse:	ø 45,5 mm; Lünette einseitig drehbar mit 60er-Teilung; Saphirglas; Krone verschraubt; Heliumventil; wasserdicht bis 60 bar (600 m)
Preis (2006):	€ 4195,-

SEAMASTER PLANET OCEAN CO-AXIAL CHRONOGRAPH

Reference number:	2918.50.38
Movement:	automatic, Omega Caliber 3313; co-axial escapement; officially certified chronometer (COSC)
Functions:	hours, minutes, subsidiary seconds; chronograph; date
Case:	stainless steel, ø 45.5 mm; unidirectionally rotating bezel with 60-minute divisions; sapphire crystal; screw-in crown; helium valve; water-resistant to 60 atm (600 m)
Price (2006):	€ 4195,-

TECHNICAL DATA
TECHNISCHE DATEN

SEAMASTER NZL-32 CHRONOMETER 150M

Referenz:	2813.30.81
Uhrwerk:	Automatik, Omega Kaliber 3602; geprüfter Chronometer
Funktionen:	Stunden, Minuten, Kleine Sekunde; Chronograph; Regattastart-Display
Gehäuse:	Edelstahl, ø 42,2 mm; Saphirglas; Krone verschraubt; wasserdicht bis 15 bar (150 m)
Preis (2006):	€ 3295,-

SEAMASTER NZL-32 CHRONOMETER 150M

Reference number:	2813.30.81
Movement:	automatic, Omega Caliber 3602; officially certified chronometer
Functions:	hours, minutes, subsidiary seconds; chronograph; regatta timing display
Case:	stainless steel, ø 42.2 mm; sapphire crystal; screw-in crown; water-resistant to 15 atm (150 m)
Price (2006):	€ 3295,-

SEAMASTER PROFESSIONAL RACING CHRONOGRAPH

Referenz:	2569.52.00
Uhrwerk:	Automatik, Omega Kaliber 3602; geprüfter Chronometer
Funktionen:	Stunden, Minuten, Kleine Sekunde; Chronograph; Regattastart-Display
Gehäuse:	Edelstahl, ø 44 m; Lünette einseitig drehbar mit 60er-Teilung; Saphirglas; wasserdicht bis 30 bar (300 m)
Preis (2002):	€ 3770,-

SEAMASTER PROFESSIONAL RACING CHRONOGRAPH

Reference number:	2569.52.00
Movement:	automatic, Omega Caliber 3602; officially certified chronometer
Functions:	hours, minutes, subsidiary seconds; chronograph; regatta timing display
Case:	stainless steel, ø 44 mm; unidirectionally rotating bezel with 60-minute divisions; sapphire crystal; water-resistant to 30 atm (300 m)
Price (2002):	€ 3770,-

SEAMASTER PLANET OCEAN CHRONO 600M CO-AXIAL

Referenz:	222.63.46.50.01.001
Uhrwerk:	Automatik, Omega Kaliber 3313; Co-Axial-Hemmung; geprüfter Chronometer (COSC)
Funktionen:	Stunden, Minuten, Kleine Sekunde; Chronograph; Datum
Gehäuse:	Rotgold, ø 45,5 mm; Lünette einseitig drehbar mit 60er-Teilung; Saphirglas; Krone und Drücker verschraubt; Heliumventil; wasserdicht bis 60 bar (600 Meter)
Preis (2008):	€ 17.620,-

SEAMASTER PLANET OCEAN CHRONO 600M CO-AXIAL

Reference number:	222.63.46.50.01.001
Movement:	automatic, Omega Caliber 3313; co-axial escapement; officially certified chronometer (COSC)
Functions:	hours, minutes, subsidiary seconds; chronograph; date
Case:	red gold, ø 45,5 mm; unidirectionally rotating bezel with 60-minute divisions; sapphire crystal; screw-in crown and buttons; helium valve; water-resistant to 60 atm (600 m)
Price (2008):	€ 17.620,-

TECHNICAL DATA
TECHNISCHE DATEN

SEAMASTER PLANET OCEAN JEWELLERY

Referenz:	222.18.42.20.01.001
Uhrwerk:	Automatik, Omega Kaliber 2500; Co-Axial-Hemmung; geprüfter Chronometer (COSC)
Funktionen:	Stunden, Minuten, Zentralsekunde; Datum
Gehäuse:	Edelstahl, ø 42 mm; Lünette einseitig drehbar mit 45 Diamanten besetzt; Saphirglas; Krone verschraubt; Heliumventil; wasserdicht bis 60 bar (600 m)
Preis (2008):	€ 9220,-

SPEEDMASTER PLANET OCEAN JEWELLERY

Reference number:	222.18.42.20.01.001
Movement:	automatic, Omega Caliber 2500; co-axial escapement; officially certified chronometer (COSC)
Functions:	hours, minutes, sweep seconds; date
Case:	stainless steel, ø 42 mm; unidirectionally rotating bezel set with 45 diamonds; sapphire crystal; screw-in crown; helium valve; water-resistant to 60 atm (600 m)
Price (2008):	€ 9220,-

SEAMASTER AQUA TERRA CO-AXIAL

Referenz:	231.10.30.20.06.001
Uhrwerk:	Automatik, Omega Kaliber 8520/8521; Co-Axial-Hemmung; geprüfter Chronometer
Funktionen:	Stunden, Minuten, Zentralsekunde; Datum
Gehäuse:	Edelstahl, ø 30 mm; Saphirglas; wasserdicht bis 15 bar (150 m)
Preis (2008):	€ 3460,-

SEAMASTER AQUA TERRA CO-AXIAL

Reference number:	231.10.30.20.06.001
Movement:	automatic, Omega Caliber 8520/8521; co-axial escapement; officially certified chronometer
Functions:	hours, minutes, sweep seconds; chronograph; date
Case:	stainless steel, ø 30 mm; sapphire crystal; water-resistant to 15 atm (150 m)
Price (2008):	€ 3460,-

SEAMASTER AQUA TERRA CO-AXIAL

Referenz:	231.53.42.21.06.001
Uhrwerk:	Automatik, Omega Kaliber 8500; Co-Axial-Hemmung; geprüfter Chronometer
Funktionen:	Stunden, Minuten, Zentralsekunde; Datum
Gehäuse:	Rotgold, ø 41,5 mm; Saphirglas; Boden mit Sichtfenster; wasserdicht bis 15 bar (150 m)
Preis (2008):	€ 10.910,-

SEAMASTER AQUA TERRA CO-AXIAL

Reference number:	231.53.42.21.06.001
Movement:	automatic, Omega Caliber 8500; co-axial escapement; officially certified chronometer
Functions:	hours, minutes, sweep seconds; date
Case:	red gold, ø 30 mm; sapphire crystal; transparent case back; water-resistant to 15 atm (150 m)
Price (2008):	€ 10.910,-

TECHNICAL DATA
TECHNISCHE DATEN

RAILMASTER CHRONOMETER

Referenz:	2802.52.37
Uhrwerk:	Automatik, Omega Kaliber 2403; Co-Axial-Hemmung; geprüfter Chronometer (COSC)
Funktionen:	Stunden, Minuten, Kleine Sekunde
Gehäuse:	Edelstahl, ø 41 mm; Saphirglas; Boden mit Sichtfenster; Krone verschraubt; wasserdicht bis 15 bar (150 m)
Preis (2003):	€ 2240,-

RAILMASTER CHRONOMETER

Reference number:	2802.52.37
Movement:	automatic, Omega Caliber 2403; co-axial escapement; officially certified chronometer (COSC)
Functions:	hours, minutes, sweep seconds
Case:	stainless steel, ø 41 mm; sapphire crystal; transparent case back; screw-in crown; water-resistant to 15 atm (150 m)
Price (2003):	€ 2240,-

SEAMASTER

SEAMASTER PLOPROF 1200M

Referenz:	224.30.55.21.01.001
Uhrwerk:	Automatik, Omega Kaliber 8500; Co-Axial-Hemmung; geprüfter Chronometer (COSC)
Funktionen:	Stunden, Minuten, Zentralsekunde; Datum
Gehäuse:	Edelstahl, ø 55 mm; beidseitig drehbare Lünette mit 60er-Teilung; Saphirglas; Krone verschraubt; Heliumventil; wasserdicht bis 120 bar (1200 m)
Preis (2009):	€ 5990,-

SEAMASTER PLOPROF 1200M

Reference number:	224.30.55.21.01.001
Movement:	automatic, Omega Caliber 8500; co-axial escapement; officially certified chronometer (COSC)
Functions:	hours, minutes, sweep seconds; date
Case:	stainless steel, ø 55 mm; unidirectionally rotating bezel with 60-minute divisions; sapphire crystal; screw-in crown; water-resistant to 120 atm (1200 m)
Price (2009):	€ 5990,-

TECHNICAL DATA
TECHNISCHE DATEN

SEAMASTER 300M GMT

Referenz:	2834.50.91
Uhrwerk:	Automatik, Omega Kaliber 1128; geprüfter Chronometer (COSC)
Funktionen:	Stunden, Minuten, Zentralsekunde; Datum; 24-Std.-Anzeige (zweite Zeitzone)
Gehäuse:	Edelstahl, ø 41,5 mm; drehbare Lünette mit 24er-Teilung; Saphirglas; Krone verschraubt; wasserdicht bis 30 bar (300 m)
Preis (1998):	€ 1740,-

SEAMASTER 300M GMT

Reference number:	2834.50.91
Movement:	automatic, Omega Caliber 1128; officially certified chronometer (COSC)
Functions:	hours, minutes, sweep seconds; date; 24-hour display (second time zone)
Case:	stainless steel, ø 41,5 mm; rotating bezel with 24-hour divisions; sapphire crystal; screw-in crown; water-resistant to 30 atm (300 m)
Price (1998):	€ 1740,-

SEAMASTER 300M APNEA

Referenz:	2895.30.91
Uhrwerk:	Automatik, Omega Kaliber 3601
Funktionen:	Stunden, Minuten, Zentralsekunde; Tauch-Chronograph mit sieben Scheiben (bis 14 Minuten)
Gehäuse:	Edelstahl, ø 41,5 mm; einseitig drehbare Lünette mit 60er-Teilung; Saphirglas, Krone und Drücker verschraubt; wasserdicht bis 30 bar (300 m)
Preis (2003):	€ 2610,-

SEAMASTER 300M APNEA

Reference number:	2895.30.91
Movement:	automatic, Omega Caliber 3601
Functions:	hours, minutes, sweep seconds; diving chronograph with seven disks (to 14 minutes)
Case:	stainless steel, ø 41.5 mm; unidirectionally rotating bezel with 60-minute divisions; sapphire crystal; screw-in crown; helium valve; water-resistant to 30 atm (300 m)
Price (2003):	€ 2610,-

TECHNICAL DATA
TECHNISCHE DATEN

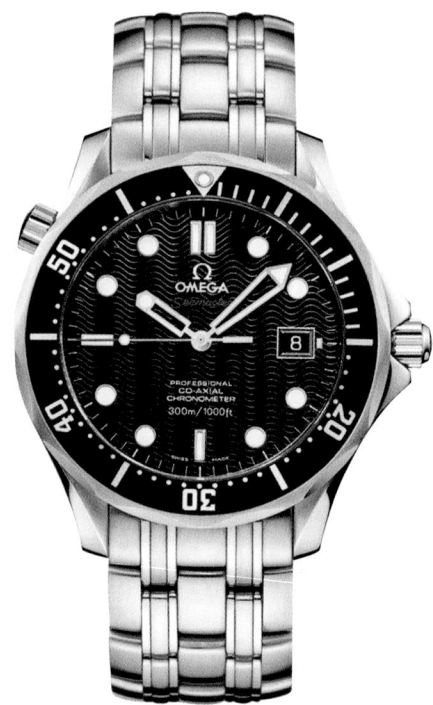

SEAMASTER DIVER CO-AXIAL

Referenz:	212.30.41.20.01.002
Uhrwerk:	Automatik, Omega Kaliber 2500; Co-Axial-Hemmung; geprüfter Chronometer (COSC)
Funktionen:	Stunden, Minuten, Zentralsekunde; Datum
Gehäuse:	Edelstahl, ø 41 mm; Lünette einseitig drehbar mit 60er-Teilung; Saphirglas; Krone verschraubt; Heliumventil; wasserdicht bis 30 bar (300 m)
Preis (2008):	€ 2460,-

SEAMASTER DIVER CO-AXIAL

Reference number:	212.30.41.20.01.002
Movement:	automatic, Omega Caliber 2500; co-axial escapement; officially certified chronometer (COSC)
Functions:	hours, minutes, sweep seconds; date
Case:	platinum, ø 41 mm; unidirectionally rotating bezel with 60-minute divisions; sapphire crystal; screw-in crown; helium valve; water-resistant to 30 atm (300 m)
Price (2008):	€ 2460,-

SEAMASTER DIVER CHRONO

Referenz:	213.30.42.0.01.001
Uhrwerk:	Automatik, Omega Kaliber 1164; geprüfter Chronometer (COSC)
Funktionen:	Stunden, Minuten, Kleine Sekunde; Chronograph; Datum
Gehäuse:	Edelstahl, ø 41,5 mm; Lünette einseitig drehbar mit 60er-Teilung; Saphirglas; Krone und Drücker verschraubt; Heliumventil; wasserdicht bis 30 bar (300 m)
Preis (2008):	€ 2610,-

SEAMASTER DIVER CHRONO

Reference number:	213.30.42.0.01.001
Movement:	automatic, Omega Caliber 1164; officially certified chronometer (COSC)
Functions:	hours, minutes, sweep seconds; date
Case:	stainless steel, ø 41,5 mm; unidirectionally rotating bezel with 60-minute divisions; sapphire crystal; screw-in crown and buttons; helium valve; water-resistant to 30 atm (300 m)
Price (2008):	€ 2610,-

SEAMASTER PLANET OCEAN CO-AXIAL

Referenz:	2900.51.82
Uhrwerk:	Automatik, Omega Kaliber 2500; Co-Axial-Hemmung; geprüfter Chronometer (COSC)
Funktionen:	Stunden, Minuten, Zentralsekunde; Datum
Gehäuse:	Edelstahl, ø 45,5 mm; Lünette einseitig drehbar mit 60er-Teilung; Saphirglas; Krone verschraubt; Heliumventil; wasserdicht bis 60 bar (600 m)
Preis (2006):	€ 2670,-

SEAMASTER PLANET OCEAN CO-AXIAL

Reference number:	2900.51.82
Movement:	automatic, Omega Caliber 2500; co-axial escapement; officially certified chronometer (COSC)
Functions:	hours, minutes, sweep seconds; date
Case:	stainless steel, ø 45.5 mm; unidirectionally rotating bezel with 60-minute divisions; sapphire crystal; screw-in crown; helium valve; water-resistant to 60 atm (600 m)
Price (2006):	€ 2670,-

TECHNICAL DATA
TECHNISCHE DATEN

SEAMASTER PLANET OCEAN CO-AXIAL 600M „ORANGE MÉCANIQUE"

Referenz:	2916.50.48
Uhrwerk:	Automatik, Omega Kaliber 3313; Co-Axial-Hemmung; geprüfter Chronometer (COSC)
Funktionen:	Stunden, Minuten, Zentralsekunde; Datum
Gehäuse:	Edelstahl, ø 45,5 mm; Lünette mit 6 orangefarbenen und 18 weißen Diamanten besetzt; Saphirglas; Krone verschraubt; Heliumventil; wasserdicht bis 60 bar (600 m)
Preis (2007):	€ 56.820,-

SEAMASTER PLANET OCEAN CO-AXIAL 600M ORANGE MÉCANIQUE

Reference number:	2916.50.48
Movement:	automatic, Omega Caliber 3313; co-axial escapement; officially certified chronometer (COSC)
Functions:	hours, minutes, sweep seconds; date
Case:	stainless steel, ø 45.5 mm; bezel set with 6 orange and 18 white diamonds; sapphire crystal; screw-in crown; helium valve; water-resistant to 60 atm (600 m)
Price (2007):	€ 56.820,-

SEAMASTER SQUELETTE 50TH ANNIVERSARY

Referenz:	2932.80.00
Uhrwerk:	Automatik, Omega Kaliber 1012; Werk von Hand skelettiert und graviert
Funktionen:	Stunden, Minuten, Zentralsekunde
Gehäuse:	Weißgold, ø 41 mm; Lünette einseitig drehbar mit 50er-Teilung; Saphirglas; Boden mit Sichtfenster; Heliumventil; wasserdicht bis 30 bar (300 m)
Preis (1998):	€ 37.610,–

SEAMASTER SQUELETTE 50TH ANNIVERSARY

Reference number:	2932.80.00
Movement:	automatic, Omega Caliber 1012; movement completely skeletonized and engraved by hand
Functions:	hours, minute, sweep seconds
Case:	white gold, ø 41 mm; unidirectionally rotating bezel with 50-minute divisions; sapphire crystal; transparent case back; helium valve; water-resistant 30 atm (300 m)
Price (1998):	€ 37.610,–

JAMES BOND – IM DIENSTE IHRER MAJESTÄT

JAMES BOND: ON HER MAJESTY'S SECRET SERVICE

Schon 1995 hat sich der bekannteste Geheimagent der Welt für Omega entschieden: In „Golden Eye" trug James Bond (Pierce Brosnan) erstmals eine Seamaster im Einsatz. Mit dem Comeback des smarten britischen Draufgängers in „Casino Royal" elf Jahre später verhalf der neue Darsteller Daniel Craig dann nicht nur der etwas angestaubten Filmreihe wieder zu Erfolgen, auch die neue Seamaster Professional Diver an seinem Handgelenk erregte einige Aufmerksamkeit. Im neuesten Streifen „Ein Quantum Trost" ist die Taucheruhr nun schon gar nicht mehr vom Handgelenk des Agenten ihrer Majestät wegzudenken – sie ist sein treuer Begleiter in allen Gefahrensituationen.

In 1995 the most famous secret agent in the world chose Omega: James Bond (Pierce Brosnan) wore a Seamaster for the first time in Golden Eye. The comeback of the clever British daredevil in Casino Royal eleven years later, saw the new Bond, Daniel Craig, not only increasing awareness for the film series, the Seamaster Professional Diver on his wrist also caused a sensation. In his newest adventure, A Quantum of Solace, the diver's watch seems inseparable from Her Majesty's favorite secret agent; it remains his loyal companion in every dangerous situation.

SEAMASTER DIVER 300 M „JAMES BOND" LIMITED EDITION

Referenz:	2226.80.00
Uhrwerk:	Automatik, Omega Kaliber 2500; Co-Axial-Hemmung; geprüfter Chronometer
Funktionen:	Stunden, Minuten, Zentralsekunde; Datum
Gehäuse:	Edelstahl, ø 41 mm; Lünette einseitig drehbar mit 60er-Teilung; Saphirglas; wasserdicht bis 30 bar (300 m)
Preis (2006):	€ 2625,-

SEAMASTER DIVER 300 M JAMES BOND LIMITED EDITION

Reference number:	2226.80.00
Movement:	automatic, Omega Caliber 2500; co-axial escapement; officially certified chronometer
Functions:	hours, minutes, sweep seconds; date
Case:	stainless steel, ø 41 mm; unidirectionally rotating bezel with 60-minute divisions; sapphire crystal; water-resistant to 30 atm (300 m)
Price (2006):	€ 2625,-

TECHNICAL DATA
TECHNISCHE DATEN

SEAMASTER DIVER 300 M „JAMES BOND"

Referenz:	2220.80.00
Uhrwerk:	Automatik, Omega Kaliber 2500; Co-Axial-Hemmung; geprüfter Chronometer
Funktionen:	Stunden, Minuten, Zentralsekunde; Datum
Gehäuse:	Edelstahl, ø 41 mm; Lünette einseitig drehbar mit 60er-Teilung; Saphirglas; wasserdicht bis 30 bar (300 m)
Preis (2006):	€ 2325,-

SEAMASTER DIVER 300 M JAMES BOND

Reference number:	2220.80.00
Movement:	automatic, Omega Caliber 2500; co-axial escapement; officially certified chronometer
Functions:	hours, minutes, sweep seconds; date
Case:	stainless steel, ø 41 mm; unidirectionally rotating bezel with 60-minute divisions; sapphire crystal; water-resistant to 30 atm (300 m)
Price (2006):	€ 2325,-

SEAMASTER PLANET OCEAN 600M „QUANTUM OF SOLACE" LIMITED EDITION

Referenz:	222.30.46.20.01.001
Uhrwerk:	Automatik, Omega Kaliber 2500; Co-Axial-Hemmung; geprüfter Chronometer
Funktionen:	Stunden, Minuten, Zentralsekunde; Datum
Gehäuse:	Edelstahl, ø 45,5 mm; Lünette einseitig drehbar mit 60er-Teilung; Saphirglas; wasserdicht bis 60 bar (600 m)
Preis (2009):	€ 3000,-

SPEEDMASTER PLANET OCEAN 600M QUANTUM OF SOLACE LIMITED EDITION

Reference number:	222.30.46.20.01.001
Movement:	automatic, Omega Caliber 2500; co-axial escapement; officially certified chronometer Functions: hours, minutes, sweep seconds; date
Case:	stainless steel, ø 45.5 mm; unidirectionally rotating bezel with 60-minute divisions; sapphire crystal; water-resistant to 60 atm (600 m)
Price (2009):	€ 3000,-

CHRONOGRAPHS
CHRONOGRAPHEN

Bieten die Speedmaster- und Seamaster-Modelle durchgehend eine in den vergangenen Jahren zunehmend im Trend liegende sportliche Optik, die Robustheit und Widerstandsfähigkeit verkörpert, zeichnen sich im Gegensatz dazu die Chronographen der Omega-Linien „De Ville" und „Constellation" durch ihre Eleganz, gepaart mit technischer Finesse, aus. Ihr Äußeres ist von klassischen Vorbildern aus der eigenen Markenhistorie inspiriert, während im Inneren mit den Co-Axial-Uhrwerken modernste Uhrmachertechnik arbeitet, die es ermöglicht, Zeiten präzise zu stoppen.

While the Speedmaster and Seamaster models have embodied the last few years' trends in sporty visuals and robustness, the chronographs of the De Ville and Constellation lines have been mainly characterized by their elegance and technical finesse. Their visuals have been inspired by the classic models of their own history, while on the inside they are outfitted with the most modern watch technology allowing time to be precisely stopped: the co-axial movement.

TECHNICAL DATA
TECHNISCHE DATEN

CONSTELLATION DOUBLE EAGLE CO-AXIAL CHRONOGRAPH

Referenz:	1819.51.91 f.c.
Uhrwerk:	Automatik, Omega Kaliber 3313; Co-Axial-Hemmung; geprüfter Chronometer (COSC)
Funktionen:	Stunden, Minuten, Kleine Sekunde; Chronograph; Datum
Gehäuse:	Edelstahl, ø 41 mm; Lünette in schwarzem Aluminium mit römischen Stundenindexen; Saphirglas; wasserdicht bis 10 bar (100 m)
Preis (2006):	€ 4080,-

CONSTELLATION DOUBLE EAGLE CO-AXIAL CHRONOGRAPH

Reference number:	1819.51.91 f.c.
Movement:	automatic, Omega Caliber 3313; co-axial escapement; officially certified chronometer (COSC)
Functions:	hours, minutes, subsidiary seconds; chronograph; date
Case:	stainless steel, ø 41mm; black aluminum bezel with Roman numeral hour markers; sapphire crystal; water-resistant to 10 atm (100 m)
Price (2006):	€ 4080,-

CONSTELLATION DOUBLE EAGLE CO-AXIAL „MISSION HILLS" CHRONOGRAPH

Referenz:	121.92.41.50.01.001
Uhrwerk:	Automatik, Omega Kaliber 3313; Co-Axial-Hemmung; geprüfter Chronometer (COSC)
Funktionen:	Stunden, Minuten, Kleine Sekunde; Chronograph; Datum
Gehäuse:	Titan, ø 41 mm; Lünette mit römischen Stundenindexen; Saphirglas; wasserdicht bis 10 bar (100 m)
Bemerkung:	Karbonfaser-Zifferblatt
Preis (2007):	€ 4890,-

CONSTELLATION DOUBLE EAGLE CO-AXIAL MISSION HILLS CHRONOGRAPH

Reference number:	121.92.41.50.01.001
Movement:	automatic, Omega Caliber 3313; co-axial escapement; officially certified chronometer (COSC)
Functions:	hours, minutes, subsidiary seconds; chronograph; date
Case:	titanium, ø 41 mm; bezel with Roman numeral hour markers; sapphire crystal; water-resistant to 10 atm (100 m)
Price (2007):	€ 4890,-

TECHNICAL DATA
TECHNISCHE DATEN

DE VILLE CO-AXIAL CHRONOSCOPE

Referenz:	4850.30.37
Uhrwerk:	Automatik, Omega Kaliber 3313; Co-Axial-Hemmung; geprüfter Chronometer (COSC)
Funktionen:	Stunden, Minuten, Kleine Sekunde; Chronograph; Datum
Gehäuse:	Edelstahl, ø 41 mm; Saphirglas
Preis (2006):	€ 4420,-

DE VILLE CO-AXIAL CHRONOSCOPE

Reference number:	4850.30.37
Movement:	automatic, Omega Caliber 3313; co-axial escapement; officially certified chronometer (COSC)
Functions:	hours, minutes, subsidiary seconds; chronograph; date
Case:	stainless steel, ø 41 mm; sapphire crystal
Price (2006):	€ 4420,-

DE VILLE CO-AXIAL CHRONOSCOPE GMT

Referenz:	422.13.44.52.13.001
Uhrwerk:	Automatik, Omega Kaliber 3603; Co-Axial-Hemmung; geprüfter Chronometer (COSC)
Funktionen:	Stunden, Minuten, Kleine Sekunde; Chronograph; Datum; 24-Std.-Anzeige (zweite Zeitzone)
Gehäuse:	Edelstahl, ø 44 mm; Saphirglas; Boden mit Sichtfenster; wasserdicht bis 10 bar (100 m)
Preis (2007):	€ 5730,-

DE VILLE CO-AXIAL CHRONOSCOPE GMT

Reference number:	422.13.44.52.13.001
Movement:	automatic, Omega Caliber 3603; co-axial escapement; officially certified chronometer (COSC)
Functions:	hours, minutes, subsidiary seconds; chronograph; date; 24-hour display (second time zone)
Case:	stainless steel, ø 44 mm; sapphire crystal; transparent case back; water-resistant to 10 atm (100 m)
Price (2007):	€ 5730,-

DE VILLE CO-AXIAL CHRONO RATTRAPANTE „PLATINUM ICE"

Uhrwerk:	Automatik, Omega Kaliber 3612; Co-Axial-Hemmung; geprüfter Chronometer (COSC)
Funktionen:	Stunden, Minuten, Kleine Sekunde; Schleppzeiger-Chronograph; Datum
Gehäuse:	Platin, ø 41 mm; Lünette mit 24 Diamanten besetzt; Saphirglas; Boden mit Sichtfenster; Krone verschraubt; wasserdicht bis 10 bar (100 m)
Preis (2006):	€ 72.990,-

DE VILLE CO-AXIAL CHRONO RATTRAPANTE PLATINUM ICE

Movement:	automatic, Omega Caliber 3612; co-axial escapement; officially certified chronometer (COSC)
Functions:	hours, minutes, subsidiary seconds; flyback chronograph; date
Case:	platinum, ø 41 mm; bezel set with 24 diamonds; sapphire crystal; transparent case back; screw-in crown; water-resistant to 10 atm (100 m)
Price (2006):	€ 72.990,-

DE VILLE CO-AXIAL CHRONO RATTRAPANTE

Referenz:	4847.50.31. f.c.
Uhrwerk:	Automatik, Omega Kaliber 3612; Co-Axial-Hemmung; geprüfter Chronometer (COSC)
Funktionen:	Stunden, Minuten, Kleine Sekunde; Schleppzeiger-Chronograph; Datum
Gehäuse:	Edelstahl, ø 41 mm; Saphirglas; Boden mit Sichtfenster; Krone verschraubt; wasserdicht bis 10 bar (100 m)
Preis (2006):	€ 8580,-

DE VILLE CO-AXIAL CHRONO RATTRAPANTE

Reference number:	4847.50.31.f.c.
Movement:	automatic, Omega Caliber 3612; co-axial escapement; officially certified chronometer (COSC)
Functions:	hours, minutes, subsidiary seconds; flyback chronograph; date
Case:	stainless steel, ø 41 mm; sapphire crystal; transparent case back; screw-in crown; water-resistant to 10 atm (100 m)
Price (2006):	€ 8580,-

TECHNICAL DATA
TECHNISCHE DATEN

DE VILLE CO-AXIAL CHRONOGRAPH
LIMITED EDITION

Referenz:	4643.20.32
Uhrwerk:	Automatik, Omega Kaliber 3313; Co-Axial-Hemmung; geprüfter Chronometer (COSC)
Funktionen:	Stunden, Minuten, Kleine Sekunde; Chronograph; Datum
Gehäuse:	Roségold, ø 41 mm; Saphirglas
Preis (2002):	€ 7730,-

DE VILLE TERRA CO-AXIAL CHRONOGRAPH
LIMITED EDITION

Reference number:	4643.20.32
Movement:	automatic, Omega Caliber 3313; co-axial escapement; officially certified chronometer (COSC)
Functions:	hours, minutes, subsidiary seconds; chronograph; date
Case:	rose gold, ø 41 mm; sapphire crystal
Price (2002):	€ 7730,-

DE VILLE X2 CO-AXIAL CHRONOGRAPH

Referenz:	423.53.37.50.01.001
Uhrwerk:	Automatik, Omega Kaliber 3202; Co-Axial-Hemmung; geprüfter Chronometer (COSC)
Funktionen:	Stunden, Minuten, Kleine Sekunde; Chronograph
Gehäuse:	Roségold, 37 x 37 mm; Saphirglas; wasserdicht bis 5 bar (50 m)
Preis (2007):	€ 12.590,-

DE VILLE X2 CO-AXIAL CHRONOGRAPH

Reference number:	423.53.37.50.01.001
Movement:	automatic, Omega Caliber 3202; co-axial escapement; officially certified chronometer (COSC)
Functions:	hours, minutes, subsidiary seconds; chronograph; date
Case:	rose gold, 37 x 37 mm; sapphire crystal; water-resistant to 5 atm (50 m)
Price (2007):	€ 12.590,-

TECHNICAL DATA
TECHNISCHE DATEN

DE VILLE CO-AXIAL CHRONOGRAPH ROME 1960

Referenz:	4841.20.32
Uhrwerk:	Automatik, Omega Kaliber 3313; Co-Axial-Hemmung; geprüfter Chronometer (COSC)
Funktionen:	Stunden, Minuten, Kleine Sekunde; Chronograph; Datum
Gehäuse:	Edelstahl, ø 41 mm; Saphirglas; wasserdicht bis 10 bar (100 m)
Preis (2004):	€ 4230,-

DE VILLE CO-AXIAL CHRONOGRAPH ROME 1960

Reference number:	4841.20.32
Movement:	automatic, Omega Caliber 3313; co-axial escapement; officially certified chronometer (COSC)
Functions:	hours, minutes, subsidiary seconds; chronograph; date
Case:	stainless steel, ø 41 mm; sapphire crystal; water-resistant to 10 atm (100 m)
Prices (2004):	€ 4230,-

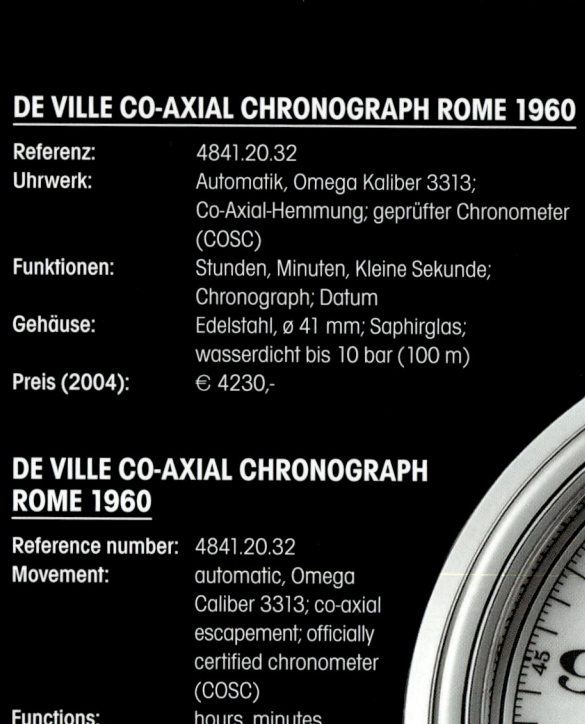

ST. MORITZ WATCH

Referenz:	4842.20.32
Uhrwerk:	Automatik, Omega Kaliber 3313; Co-Axial-Hemmung; geprüfter Chronometer (COSC)
Funktionen:	Stunden, Minuten, Kleine Sekunde; Chronograph; Datum
Gehäuse:	Edelstahl, ø 41 mm; Saphirglas; wasserdicht bis 10 bar (100 m)
Preis (2004):	€ 4200,-

ST. MORITZ WATCH

Reference number:	4842.20.32
Movement:	automatic, Omega Caliber 3313; co-axial escapement; officially certified chronometer (COSC)
Functions:	hours, minutes, subsidiary seconds; chronograph; date
Case:	stainless steel, ø 41 mm; sapphire crystal; water-resistant to 10 atm (100 m)
Price (2004):	€ 4200,-

TECHNICAL DATA
TECHNISCHE DATEN

MUSEUM COLLECTION „RACEND TIMER"

Referenz:	516.53.39.50.02.001
Uhrwerk:	Automatik, Omega Kaliber 3201; Co-Axial-Hemmung; geprüfter Chronometer (COSC)
Funktionen:	Stunden, Minuten, Kleine Sekunde; Chronograph
Gehäuse:	Roségold, ø 39 mm; Saphirglas; wasserdicht bis 3 bar (30 m)
Preis (2008):	€ 10.370,-

MUSEUM COLLECTION RACEND TIMER

Reference number:	516.53.39.50.02.001
Movement:	automatic, Omega Caliber 3201; co-axial escapement; officially certified chronometer (COSC)
Functions:	hours, minutes, subsidiary seconds; chronograph
Case:	rose gold, ø 39 mm; sapphire crystal; water-resistant to 3 atm (30 m)
Price (2008):	€ 10.370,-

OMEGA POCKET WATCH 1932

Referenz:	5110.20.00
Uhrwerk:	Handaufzug, Omega Kaliber 3889A
Funktionen:	Stunden, Minuten, Kleine Sekunde; Schleppzeiger-Chronograph
Gehäuse:	Weißgold, ø 57 mm; Saphirglas; Boden mit Sprungdeckel und Sichtfenster
Preis (2008):	€ 67.720,-

OMEGA POCKET WATCH 1932

Reference number:	5110.20.00
Movement:	manually wound, Omega Caliber 3889A
Functions:	hours, minutes, subsidiary seconds; split-seconds chronograph
Case:	white gold, ø 57 mm; sapphire crystal; transparent case back with hinged lid
Price (2008):	€ 67.720,-

OMEGA
IM ZEICHEN DER RINGE –
OMEGA UND OLYMPIA

Omega ist seit 1932 eng mit den Olympischen Spielen verbunden. Seit die Schweizer Marke damals in Los Angeles zum ersten Mal für die offizielle Zeitmessung verantwortlich zeichnete, ist eine enge Verbundenheit zu dem alle vier Jahre stattfindenden wichtigsten Ereignis des internationalen Sports erwachsen. Anlässlich der letzten Olympischen Sommerspiele in Peking 2008 legte Omega gleich zwei Olympia-Armbanduhrenkollektionen vor: Zum einen die „Beijing 2008 Olympic Collection", die bereits von Sommer 2007 an nach und nach publikumswirksam lanciert wurde, und zum anderen die „Timeless Collection", die sich als eine Hommage an die Tradition der Omega-Zeitmessung begreift. Und kaum sind die Spiele in Peking vorbei, werfen die Winterspiele in Vancouver 2010 ihren Schatten voraus...

UNDER THE SIGN OF THE RINGS:
OMEGA AND THE OLYMPICS

Omega has been closely tied to the Olympic Games since 1932. Since the Swiss brand became responsible for the official timing in Los Angeles in that year, a close tie to the most important event in international sports has grown between them. Omega came out with two Olympic wristwatch collections on the occasion of the most recent Olympic Summer Games in Beijing in 2008: the Beijing 2008 Olympic Collection, which was successively launched in the summer of 2007, and the Timeless Collection, which represents an homage to the tradition of Omega's timing. The games in Beijing had hardly closed before Vancouver's 2010 Winter Games began casting their shadow...

TECHNICAL DATA
TECHNISCHE DATEN

**SPEEDMASTER CO-AXIAL
„FIVE-COUNTER-CHRONOGRAPH"**

Referenz:	321.53.44.52.01.001
Uhrwerk:	Automatik, Omega Kaliber 3888; Co-Axial-Hemmung; geprüfter Chronometer
Funktionen:	Stunden, Minuten, Kleine Sekunde; Chronograph; Datum; Wochentag
Gehäuse:	Rotgold, ø 44,25 mm; Lünette mit Tachymeterskala; Saphirglas; wasserdicht bis 10 bar (100m)
Preis (2008):	€ 14.590,-

**SPEEDMASTER CO-AXIAL
FIVE-COUNTER-CHRONOGRAPH**

Reference number:	321.53.44.52.01.001
Movement:	automatic, Omega Caliber 3888; co-axial escapement; officially certified chronometer
Functions:	hours, minutes, subsidiary seconds; chronograph; date; weekday
Case:	red gold, ø 44.25 mm; bezel with tachymeter; sapphire crystal; water-resistant to 10 atm (100 m)
Price (2008):	€ 14.590,-

TECHNICAL DATA
TECHNISCHE DATEN

SEAMASTER XXIX

Referenz:	516.53.37.20.09.008
Uhrwerk:	Automatik, Omega Kaliber 2403; Co-Axial-Hemmung; geprüfter Chronometer
Funktionen:	Stunden, Minuten, Zentralsekunde
Gehäuse:	Gelbgold, ø 37 mm; Saphirglas; wasserdicht bis 10 bar (100 m)
Preis (2008):	€ 7680,-

SEAMASTER XXIX

Reference number:	516.53.37.20.09.008
Movement:	automatic, Omega Caliber 2403; co-axial escapement; officially certified chronometer
Functions:	hours, minutes, sweep seconds
Case:	yellow gold, ø 37 mm; sapphire crystal; water-resistant to 10 atm (100 m)
Price (2008):	€ 7680,-

SEAMASTER „AQUA TERRA"

Referenz: 221.20.42.40.01.001 und 221.10.42.40.01.001
Uhrwerk: Automatik, Omega Kaliber 3301; geprüfter Chronometer
Funktionen: Stunden, Minuten, Kleine Sekunde; Chronograph; Datum
Gehäuse: Edelstahl/Edelstahl mit Rotgold-Lünette, ø 42,2 mm; Saphirglas; wasserdicht bis 15 bar (150 m)
Preis (2008): € 3840,- in Edelstahl, € 5300,- mit Rotgold-Lünette

SEAMASTER AQUA TERRA

Reference number: 221.20.42.40.01.001 and 221.10.42.40.01.001
Movement: automatic, Omega Caliber 3301; officially certified chronometer
Functions: hours, minutes, subsidiary seconds; chronograph; date
Case: stainless steel/stainless steel with red gold bezel, ø 42,2 mm; sapphire crystal; water-resistant to 15 atm (150 m)
Price (2008): € 3840,- stainless steel, € 5300,- with red gold bezel

UNDER THE SIGN OF THE RINGS

DE VILLE CHRONOSCOPE

Referenz:	422.13.41.50.04.001
Uhrwerk:	Automatik, Omega Kaliber 3313; Co-Axial-Hemmung; geprüfter Chronometer
Funktionen:	Stunden, Minuten, Kleine Sekunde; Chronograph
Gehäuse:	ø 41 mm; Saphirglas; wasserdicht bis 10 bar (100 m)
Preis (2008):	€ 4530,-

DE VILLE CHRONOSCOPE

Reference number:	422.13.41.50.04.001
Movement:	automatic, Omega Caliber 3313; co-axial escapement; officially certified chronometer
Functions:	hours, minutes, subsidiary seconds; chronograph
Case:	stainless steel, ø 41 mm; sapphire crystal; water-resistant to 10 atm (100 m)
Price (2008):	€ 4530,-

SPEEDMASTER „BROAD ARROW"

Referenz:	321.10.42.50.04.001
Uhrwerk:	Automatik, Omega Kaliber 3313; Co-Axial-Hemmung; geprüfter Chronometer
Funktionen:	Stunden, Minuten, Kleine Sekunde; Chronograph; Datum
Gehäuse:	Edelstahl, ø 42 mm; Lünette mit Tachymeterskala; Saphirglas; wasserdicht bis 10 bar (100m)
Preis (2008):	€ 4380,-

SPEEDMASTER BROAD ARROW

Reference number:	321.10.42.50.04.001
Movement:	automatic, Omega Caliber 3313; co-axial escapement; officially certified chronometer
Functions:	hours, minutes, subsidiary seconds; chronograph; date
Case:	stainless steel, ø 42 mm; bezel with tachymeter; sapphire crystal; water-resistant to 10 atm (100 m)
Price (2008):	€ 4380,-

TECHNICAL DATA
TECHNISCHE DATEN

SEAMASTER PLANET OCEAN

Referenz:	222.32.46.50.01.001
Uhrwerk:	Automatik, Omega Kaliber 3313; Co-Axial-Hemmung; geprüfter Chronometer
Funktionen:	Stunden, Minuten, Kleine Sekunde; Chronograph; Datum
Gehäuse:	Edelstahl, ø 45,5 mm; Lünette einseitig drehbar mit 60er-Teilung; Saphirglas; wasserdicht bis 60 bar (600 m)
Preis (2008):	€ 4530,-

SEAMASTER PLANET OCEAN

Reference number:	222.32.46.50.01.001
Movement:	automatic, Omega Caliber 3313; co-axial escapement; officially certified chronometer
Functions:	hours, minutes, subsidiary seconds; chronograph; date
Case:	stainless steel, ø 45.5 mm; unidirectionally rotating bezel with 60-minute divisions; sapphire crystal; water-resistant to 60 atm (600 m)
Price (2008):	€ 4530,-

UNDER THE SIGN OF THE RINGS

TECHNICAL DATA
TECHNISCHE DATEN

CONSTELLATION LIMITED EDITION LADIES

Referenz:	111.50.36.10.52.002
Uhrwerk:	quarzgesteuert, Omega Kaliber 1456
Funktionen:	Stunden, Minuten
Gehäuse:	Rotgold, ø 22,5 mm; Lünette mit 30 Diamanten besetzt; Saphirglas; wasserdicht bis 3 bar (30 m)
Bemerkung:	Zifferblatt mit 12 Diamant-Indexen
Preis (2008):	€ 10.140,-

CONSTELLATION LIMITED EDITION LADIES

Reference number:	110.50.36.10.52.002
Movement:	quartz, Omega Caliber 1456
Functions:	hours, minutes
Case:	red gold, ø 22.5 mm; bezel set with 30 diamonds; sapphire crystal; water-resistant to 3 atm (30 m)
Remarks:	dial set with 12 diamond markers
Price (2008):	€ 10.140,-

CONSTELLATION LIMITED EDITION GENTS

Referenz:	111.55.23.60.55.002
Uhrwerk:	Omega Kaliber 1120; geprüfter Chronometer
Funktionen:	Stunden, Minuten, Zentralsekunde; Datum
Gehäuse:	Rotgold, ø 35,5 mm; Lünette mit römischen Ziffern graviert; Saphirglas; wasserdicht bis 5 bar (50 m)
Bemerkung:	Zifferblatt mit elf Diamant-Indexen
Preis (2008):	€ 7380,-

CONSTELLATION LIMITED EDITION GENTS

Reference number:	111.55.23.60.55.002
Movement:	Omega Caliber 1120; officially certified chronometer
Functions:	hours, minutes, sweep seconds; date
Case:	red gold, ø 35.5 mm; bezel engraved with Roman numerals; sapphire crystal; water-resistant to 5 atm (50 m)
Remarks:	dial set with 11 diamond markers
Price (2008):	€ 7380,-

OMEGA SEAMASTER DIVER 300M „VANCOUVER 2010" LIMITED EDITION

Referenz:	2569.52.00
Werk:	Automatik, Omega Kaliber 2500; Co-Axial-Hemmung;
Funktionen:	Stunden, Minuten, Zentralsekunde; Datum
Gehäuse:	Edelstahl, ø 41 mm; Lünette einseitig drehbar mit 60er-Teilung; Saphirglas; wasserdicht bis 30 bar (300 m)
Preis (2009):	€ 2770,-

OMEGA SEAMASTER DIVER 300M VANCOUVER 2010 LIMITED EDITION

Reference number:	2569.52.00
Movement:	automatic, Omega Caliber 2500; co-axial escapement
Functions:	hours, minutes, sweep seconds; date
Case:	stainless steel, ø 41 mm; unidirectionally rotating bezel with 60-minute divisions; sapphire crystal; water-resistant to 30 atm (300 m)
Price (2009):	€ 2770,-

OMEGA
ELEGANTE UHREN

Robustheit und Genauigkeit sind seit jeher die Markenzeichen von Omega. Diese Tradition ist über Jahrzehnte entstanden und lässt sich nicht einfach kreieren oder gar kaufen. In den vergangenen 15 Jahren hat es die Marke aber auch verstanden, ganz andere Werte, die in der Markengeschichte ebenso verankert sind, wieder zum Leben zu erwecken: Luxuriöse Eleganz und höchste uhrmacherische Fertigkeiten. Neben den seit 2001 neu aufgelegten Modellen der „Museumskollektion", die legendären Uhren aus der eigenen Geschichte huldigt, ist Omega 1994 wieder in die Welt der „Haute Horlogerie" (Hohe Uhrmacherei) vorgestoßen. Seitdem befindet sich ein Zentraltourbillon in der Kollektion, das in seiner neuesten Entwicklungsstufe seit 2007 mit Co-Axial-Hemmung ausgestattet ist, und somit seinesgleichen in der Uhrenbranche sucht.

ELEGANT WATCHES

Robustness and precision have always been the defining characteristics of Omega; such a tradition cannot simply be created or even bought over the course of decades. In the past fifteen years, the brand has, however, begun to understand and resuscitate other values that are also rooted in the brand's history: luxurious elegance and the highest horological skill. Alongside the new models of the Museum Collection that have debuted since 2001 and pay homage to the legendary watches of the brand's own history, Omega also reentered the world of haute horlogerie — the highest instance of the watchmaker's art — in 1994. At this time, a Central Tourbillon was added to the collection, which in its newest incarnation presented in 2007, is outfitted with a co-axial escapement and is thus without equal in the watch industry.

TECHNICAL DATA
TECHNISCHE DATEN

DE VILLE TOURBILLON CO-AXIAL CHRONOMETER

Referenz:	513.53.39.21.99.001
Uhrwerk:	Automatik, Omega Kaliber 2635; Co-Axial-Hemmung; Zentral-Tourbillon mit Käfig aus Titan; geprüfter Chronometer (COSC)
Funktionen:	Stunden, Minuten („Mysterieuse"-Anzeige ohne sichtbare Zeigerachse), Zentralsekunde (am Tourbillon)
Gehäuse:	Roségold, ø 38,7 mm; Saphirglas
Preis (2007):	€ 83.910,-

DE VILLE TOURBILLON CO-AXIAL CHRONOMETER

Reference number:	513.53.39.21.99.001
Movement:	automatic, Omega Caliber 2635; co-axial escapement; central tourbillon with titanium cage; officially certified chronometer (COSC)
Functions:	hours, minutes ("mysterious" display without visible hand arbors), sweep seconds (on tourbillon)
Case:	rose gold, ø 38.7 mm; sapphire crystal
Price (2007):	€ 83.910,-

ELEGANT WATCHES

TECHNICAL DATA
TECHNISCHE DATEN

DE VILLE TOURBILLON

Referenz:	5113.30.00
Uhrwerk:	Automatik, Omega Kaliber 2600; Zentral-Tourbillon
Funktionen:	Stunden, Minuten; Zentralsekunde (am Tourbillon)
Gehäuse:	Gelbgold, ø 38,7 mm; Saphirglas; wasserdicht bis 3 bar (30 m)
Preis (1994):	€ 70.739,-

DE VILLE TOURBILLON

Reference number:	5113.30.00
Movement:	automatic, Omega Caliber 2600; central tourbillon
Functions:	hours, minutes, sweep seconds (on tourbillon)
Case:	yellow gold, ø 38,7 mm; sapphire crystal; water-resistant to 3 atm (30 m)
Price (1994):	€ 70.739,-

DE VILLE HOUR VISION ANNUAL CALENDAR

Referenz:	431.63.41.22.01.001
Uhrwerk:	Automatik, Omega Kaliber 8611; Co-Axial-Hemmung; geprüfter Chronometer (COSC)
Funktionen:	Stunden, Minuten, Zentralsekunde; Jahreskalender mit Datum und Monat
Gehäuse:	Roségold, ø 41 mm; seitliche Sichtfenster; Saphirglas; Boden mit Sichtfenster; wasserdicht bis 10 bar (100 m)
Preis (2008):	€ 13.940,-

DE VILLE HOUR VISION ANNUAL CALENDAR

Reference number:	431.63.41.22.01.001
Movement:	automatic, Omega Caliber 8611; co-axial escapement; officially certified chronometer (COSC)
Functions:	hours, minutes, sweep seconds; annual calend with date and month
Case:	rose gold, ø 41 mm; transparent edges; sapphir crystal; transparent case back; water-resistant to 10 atm (100 m)
Price (2008):	€ 13.940,-

DE VILLE CO-AXIAL

Referenz:	4632.31.31
Uhrwerk:	Automatik, Omega Kaliber 2627; Co-Axial-Hemmung; geprüfter Chronometer (COSC)
Funktionen:	Stunden, Minuten, Kleine Sekunde; Datum; Gangreserveanzeige
Gehäuse:	Gelbgold, ø 38,7 mm; Saphirglas; wasserdicht bis 10 bar (100 m)
Preis (2007):	€ 7610,-

DE VILLE CO-AXIAL CHRONOSCOPE GMT

Reference number:	4632.31.31
Movement:	automatic, Omega Caliber 2627; co-axial escapement; officially certified chronometer (COSC)
Functions:	hours, minutes, subsidiary seconds; chronograph; date; power reserve display
Case:	yellow gold, ø 38.7 mm; sapphire crystal; water-resistant to 10 atm (100 m)
Price (2007):	€ 7610,-

TECHNICAL DATA
TECHNISCHE DATEN

COLLECTOR'S PIECE NO. 1
(MUSEUM COLLECTION 2001)

Referenz:	5700.50.07
Uhrwerk:	Automatik, Omega Kaliber 2200
Funktionen:	Stunden, Minuten, Kleine Sekunde
Gehäuse:	Edelstahl, ø 40,5 mm; beidseitig drehbare Lünette; Saphirglas
Preis (2001):	€ 3550,-

COLLECTOR'S PIECE NO.1
(MUSEUM COLLECTION 2001)

Reference number:	5700.50.07
Movement:	automatic, Omega Caliber 2200
Functions:	hours, minutes, subsidiary seconds
Case:	stainless steel, ø 40,5 mm; bidirectionally rotating bezel; sapphire crystal
Price (2001):	€ 3550,-

COLLECTOR'S PIECE NO. 2 (MUSEUM COLLECTION 2002)

Referenz:	5701.80.03
Uhrwerk:	Automatik, Omega Kaliber 2601
Funktionen:	Stunden, Minuten, Kleine Sekunde; Datum, Wochentag, Monat, Mondphasen
Gehäuse:	Roségold, 33,4 x 33,4 mm; Saphirglas; wasserdicht bis 5 bar (50 m)
Preis (2002):	€ 8590,-

COLLECTOR'S PIECE NO. 2 (MUSEUM COLLECTION 2002)

Reference number:	5701.80.03
Movement:	automatic, Omega Caliber 2601
Functions:	hours, minutes, subsidiary seconds; date, weekday, month, moon phase
Case:	rose gold, 33,4 x 33,4 mm; sapphire crystal; water-resistant to 5 atm (50 m)
Price (2002):	€ 8590,-

ELEGANT WATCHES

TECHNICAL DATA
TECHNISCHE DATEN

**COLLECTOR'S PIECE NO. 3
(MUSEUM COLLECTION 2003)**

Referenz:	5702.50.02
Uhrwerk:	Handaufzug, Omega Kaliber 3200
Funktionen:	Stunden, Minuten, Kleine Sekunde; Chronograph
Gehäuse:	Edelstahl, ø 38,4 mm; Lünette mit arabischen Ziffern; Saphirglas;
Preis (2003):	€ 3420,-

COLLECTOR'S PIECE NO. 3 (MUSEUM COLLECTION 2003)

Reference number:	5702.50.02
Movement:	manually wound, Omega Caliber 3200
Functions:	hours, minutes, subsidiary seconds; chronograph
Case:	stainless steel, ø 38.4 mm; bezel with Arabic numerals; sapphire crysta
Price (2003):	€ 3420,-

**COLLECTOR'S PIECE NO. 4
(MUSEUM COLLECTION 2004)**

Referenz:	5703.30.01
Uhrwerk:	Automatik, Omega Kaliber 2200 B6
Funktionen:	Stunden, Minuten, Kleine Sekunde
Gehäuse:	Roségold, ø 43,5 x 32,5 mm; Saphirglas; wasserdicht bis 3 bar (30 m)
Preis (2004):	€ 6270,-

**COLLECTOR'S PIECE NO. 4
(MUSEUM COLLECTION 2004)**

Reference number:	5703.30.01
Movement:	automatic, Omega Caliber 2200 B6
Functions:	hours, minutes, subsidiary seconds
Case:	rose gold, ø 43,5 x 32,5 mm; sapphire crystal; water-resistant to 3 atm (30 m)
Price (2004):	€ 6270,-

TONNEAU RENVERSÉ
(MUSEUM COLLECTION 2006)

Referenz:	5705.30.01
Uhrwerk:	Automatik, Omega Kaliber 2202; Co-Axial-Hemmung; geprüfter Chronometer (COSC)
Funktionen:	Stunden, Minuten, Kleine Sekunde
Gehäuse:	Weißgold/Roségold, 35,8 x 46,2 mm; Saphirglas; wasserdicht bis 3 bar (30 m)
Preis (2006):	€ 9780,-

TONNEAU RENVERSÉ
(MUSEUM COLLECTION 2006)

Reference number:	5705.30.01
Movement:	automatic, Omega Caliber 2202; co-axial escapement; officially certified chronometer (COSC)
Functions:	hours, minutes, subsidiary seconds
Case:	white gold/rose gold, 35.8 x 46.2 mm; sapphire crystal; water-resistant to 3 atm (30 m)
Price (2006):	€ 9780,-

MARINE 1932 (MUSEUM COLLECTION 2007)

Referenz:	516.53.32.20.02.002
Uhrwerk:	Handaufzug, Omega Kaliber 2007; Co-Axial-Hemmung
Funktionen:	Stunden, Minuten
Gehäuse:	Roségold/Weißgold (Innengehäuse), 33,05 x 50,5 mm; Saphirglas; Boden mit Sichtfenster; wasserdicht bis 13 bar (130 m)
Preis (2007):	€ 26.880,-

MARINE 1932 (MUSEUM COLLECTION 2007)

Reference number:	516.53.32.20.02.002
Movement:	manually wound, Omega Caliber 2007; co-axial escapement
Functions:	hours, minutes
Case:	rose gold/white gold (inner case), 33.05 x 50.5 mm; sapphire crystal; transparent case back; water-resistant to 13 atm (130 m)
Price (2007):	€ 26.880,-

ELEGANT WATCHES

TECHNICAL DATA
TECHNISCHE DATEN

CONSTELLATION '95 „OMEGA 160 YEARS"

Referenz	111.55.36.20.52.001 / 111.55.36.20.58.001 / 111.25.36.20.52.001
Uhrwerk:	Automatik, Omega Kaliber 2500; Co-Axial-Hemmung; geprüfter Chronometer (COSC)
Funktionen:	Stunden, Minuten, Zentralsekunde; Datum
Gehäuse:	Gelbgold/Roségold/Weißgold; ø 35,5 mm; Lünette mit römischen Ziffern graviert bzw. mit 40 Diamanten besetzt; Saphirglas; wasserdicht bis 5 bar (50 m)
Preis (2008):	€ 14.320,-/€ 14.320,-/€ 7300,-

CONSTELLATION '95 OMEGA 160 YEARS

Reference number:	111.55.36.20.52.001 / 111.55.36.20.58.001 / 111.25.36.20.52.001
Movement:	automatic, Omega Caliber 2500; co-axial escapement; officially certified chronometer (COSC)
Functions:	hours, minutes, sweep seconds; date
Case:	yellow gold/rose gold/white gold, ø 35.5 mm; bezel engraved with Roman numerals set with 40 diamonds; sapphire crystal; water-resistant to 5 atm (50 m)
Price (2008):	€ 14.320,-/€ 14.320,-/€ 7300,-

CONSTELLATION CHRONOMETER

Referenz:	123.10.38.21.01.001
Uhrwerk:	Automatik, Omega Kaliber 8500; Co-Axial-Hemmung; geprüfter Chronometer (COSC)
Funktionen:	Stunden, Minuten, Zentralsekunde; Datum
Gehäuse:	Weißgold, ø 38 mm; Lünette mit römischen Ziffern graviert; Saphirglas, Boden mit Sichtfenster; wasserdicht bis 10 bar (10 m)
Preis (2009):	€ 3760,-

CONSTELLATION CHRONOMETER

Reference number:	123.10.38.21.01.001
Movement:	automatic, Omega Caliber 8500; co-axial escapement; officially certified chronometer (COSC)
Functions:	hours, minutes, sweep seconds; date
Case:	white gold, ø 38 mm; bezel engraved with Roman numerals; sapphire crystal; transparent case back; water-resistant to 10 atm (10 m)
Price (2009):	€ 3760,-

TECHNISCHE DATEN
TECHNICAL DATA

DE VILLE X2 CO-AXIAL BIG DATE

Referenz:	7711.30.39 f.c.
Uhrwerk:	Automatik, Omega Kaliber 2610; Co-Axial-Hemmung; geprüfter Chronometer (COSC)
Funktionen:	Stunden, Minuten, Zentralsekunde; Großdatum
Gehäuse:	Weißgold, 35 x 35 mm; Saphirglas; wasserdicht bis 5 bar (50 m)
Preis (2006):	€ 9640,-

DE VILLE X2 CO-AXIAL BIG DATE

Reference number:	7711.30.39 f.c.
Movement:	automatic, Omega Caliber 2610; co-axial escapement; officially certified chronometer (COSC)
Functions:	hours, minutes, sweep seconds; large date
Case:	white gold, 35 x 35 mm; sapphire crystal; water-resistant to 5 atm (50 m)
Price (2006):	€ 9640,-

DE VILLE PRESTIGE CO-AXIAL SMALL SECONDS

Referenz:	4813.40.01
Uhrwerk:	Automatik, Omega Kaliber 2202; Co-Axial-Hemmung; geprüfter Chronometer (COSC)
Funktionen:	Stunden, Minuten, Kleine Sekunde
Gehäuse:	Edelstahl, ø 39 mm; Saphirglas; wasserdicht bis 3 bar (30 m)
Preis (2006):	€ 2360,-

DE VILLE PRESTIGE CO-AXIAL SMALL SECONDS

Reference number:	4813.40.01
Movement:	automatic, Omega Caliber 2202; co-axial escapement; officially certified chronometer (COSC)
Functions:	hours, minutes, subsidiary seconds
Case:	stainless steel, ø 39 mm; sapphire crystal; water-resistant to 3 atm (30 m)
Price (2006):	€ 2360,-

DE VILLE „HOUR VISION"

Referenz:	431.63.41.21.02.001
Uhrwerk:	Automatik, Omega Kaliber 8501; Co-Axial-Hemmung; geprüfter Chronometer (COSC)
Funktionen:	Stunden, Minuten, Zentralsekunde; Datum
Gehäuse:	Roségold, ø 41 mm; seitlich verglast; Saphirglas; Boden mit Sichtfenster; wasserdicht bis 10 bar (100 m)
Preis (2006):	€ 10.610,-

DE VILLE HOUR VISION

Reference number:	431.63.41.21.02.001
Movement:	automatic, Omega Caliber 8501; co-axial escapement; officially certified chronometer (COSC)
Functions:	hours, minutes, sweep seconds; date
Case:	rose gold, ø 41 mm; crystal sides; sapphire crystal; transparent case back; water-resistant to 10 atm (100 m)
Price (2006):	€ 10.610,-

TECHNICAL DATA
TECHNISCHE DATEN

CONSTELLATION '09

Referenz:	123.20.35.20.02.002
Uhrwerk:	Automatik, Omega Kaliber 2500; Co-Axial-Hemmung; geprüfter Chronometer (COSC)
Funktionen:	Stunden, Minuten, Zentralsekunde; Datum
Gehäuse:	Edelstahl, ø 35 mm; Lünette in Gelbgold mit römischen Ziffern graviert; Saphirglas; Boden mit Sichtfenster; Krone verschraubt; wasserdicht bis 10 bar (100 m)
Preis (2009):	€ 4220,-

CONSTELLATION '09

Reference number:	123.20.35.20.02.002
Movement:	automatic, Omega Caliber 2500; co-axial escapement; officially certified chronometer (COSC)
Functions:	hours, minutes, sweep seconds; date
Case:	stainless steel, ø 35 mm; yellow gold bezel engraved with Roman numerals; sapphire crystal; transparent case back; water-resistant to 10 atm (100 m)
Price (2009):	€ 4220,-

CONSTELLATION '09

Referenz:	123.10.31.20.01.001
Uhrwerk:	Automatik, Omega Kaliber 8520/8521; Co-Axial-Hemmung; geprüfter Chronometer (COSC)
Funktionen:	Stunden, Minuten, Zentralsekunde; Datum
Gehäuse:	Edelstahl, ø 31 mm; Lünette mit römischen Ziffern graviert; Saphirglas; Boden mit Sichtfenster; Krone verschraubt; wasserdicht bis 10 bar (100 m)
Preis (2009):	€ 3760,-

CONSTELLATION '09

Reference number:	123.10.31.20.01.001
Movement:	automatic, Omega Caliber 8520/8521; co-axial escapement; officially certified chronometer (COSC)
Functions:	hours, minutes, sweep seconds; date
Case:	stainless steel, ø 31 mm; bezel engraved with Roman numerals; sapphire crystal; transparent case back; water-resistant to 10 atm (100 m)
Price (2009):	€ 3760,-

CONSTELLATION '09

Referenz:	123.25.31.20.55.002
Uhrwerk:	Automatik, Omega Kaliber 8520/8521; Co-Axial-Hemmung; geprüfter Chronometer (COSC)
Funktionen:	Stunden, Minuten, Zentralsekunde; Datum
Gehäuse:	Edelstahl, ø 31 mm; Lünette in Gelbgold mit 24 Diamanten besetzt; Saphirglas; Boden mit Sichtfenster; Krone verschraubt; wasserdicht bis 10 bar (100 m)
Preis (2009):	€ 8100,-

CONSTELLATION '09

Reference number:	123.25.31.20.55.002
Movement:	automatic, Omega Caliber 8520/8521; co-axial escapement; officially certified chronometer (COSC)
Functions:	hours, minutes, sweep seconds; date
Case:	stainless steel, ø 31 mm; yellow gold bezel set with 24 diamonds; sapphire crystal; transparent case back; screw-down crown; water-resistant to 10 atm (100 m)
Price (2009):	€ 8100,-

ELEGANT WATCHES

Exklusive Armbanduhren
Ein perfektes Geschenkbuch für Freunde exklusiver Armbanduhren! Dieser kompakte Bildband zeigt Hunderte brillanter, exklusiv geschossener Fotos faszinierender Luxus-Armbanduhren.
432 Seiten, 600 farb. Abb.,
210 x 172 mm, geb.
ISBN 978-3-86852-021-7
€ (D) 19,95

Peter Braun (Hrsg.)
ARMBANDUHREN Katalog 2010
Die aktualisierte Marktübersicht zum Angebot an hochwertigen Armbanduhren enthält über 1200 Uhren mit allen Spezifikationen, Preisangaben und Bildern.
432 Seiten, 1300 farb. Abb.,
210 x 297 mm, kartoniert
ISBN 978-3-86852-065-1
€ (D) 19,90

Henning Mützlitz
A. Lange & Söhne Highlights
Dieses mit opulenten Bildern und informativem Text ausgestattete Buch präsentiert die schönsten Modelle der Nobelmarke mit Sitz in Glashütte, einem traditionsreichen Zentrum der deutschen Uhrenindustrie.
96 Seiten, ca. 150 farb. Abb., 210 x 225 mm, geb.
ISBN 978-3-86852-231-0
ca. € (D) 12,95

Henning Mützlitz
Breitling Highlights
Keine Armbanduhrenmarke produziert so charakteristisch maskuline Uhren wie Breitling. Das macht sie seit den dreißiger Jahren des letzten Jahrhunderts so unverwechselbar und beliebt.
96 Seiten, ca. 150 farb. Abb., 210 x 225 mm, geb.
ISBN: 978-3-86852-197-9
ca. € (D) 12,95

Christoph Heel (Hrsg.)
Rolex
Seit 15 Jahren begleitet das Fachmagazin ArmbandUhren die Kultmarke Rolex mit ihren exklusiven Produkten. Dieses Special präsentiert die spannendsten Stories und besten Bilder.
144 Seiten, ca. 350 farb. Abb., 210 x 297 mm, geb.
ISBN 978-3-86852-189-4
€ (D) 19,95

Stefan Muser / Michael Ph. Horlbeck
ARMBANDUHREN Klassik Katalog
Preisguide für Uhrensammler
Fundierte Aussagen zu über 1300 Modellen machen die aktuelle Ausgabe zu einem wertvollen Nachschlagewerk für Sammler seltener Stücke.
240 Seiten, ca. 1000 Abb.,
210 x 207 mm, kart.
ISBN 978-3-86852-063-7
€ (D) 19,90

Michael Ph. Horlbecks
Lexikon der Uhrenmarken
In akribischer Recherchearbeit hat Uhrenspezialist Michael Horlbeck das kaum durchdringliche Dickicht um bekannte Marken und deren Unterlabels im Bereich wertiger Armbanduhren transparent gemacht.
304 Seiten, 100 Abb., 175 x 245 mm, geb.
ISBN 978-3-86852-022-4
€ (D) 29,90

Unser komplettes Programm finden Sie unter: **www.heel-verlag.de**
Telefon: 02223 9230-38 · Fax: 02223 9230-13
E-Mail: s.becker@heel-verlag.de

HEEL Verlag GmbH
Gut Pottscheidt
53639 Königswinter